NATIVES OF A DRY PLACE

NATIVES

STORIES OF DAKOTA BEFORE THE OIL BOOM

OF A DRY PLACE

RICHARD EDWARDS

SOUTH DAKOTA HISTORICAL SOCIETY PRESS

Pierre

This publication is funded, in part, by the
Great Plains Education Foundation, Inc., Aberdeen, S.Dak.,
and by the University of Nebraska.

Library of Congress Control Number: 2015947923

The paper in this book meets the guidelines for
permanence and durability of the Committee on Production Guidelines
for Book Longevity of the Council on Library Resources.

Cover image: Main Street, Stanley, North Dakota, 1910.
State Historical Society of North Dakota

Frontispiece: The east side of Main Street in Stanley, North Dakota, 1910.
Richard Edwards collection

Maps by Michael Cooper and Katie Nieland

Text and cover design by Rich Hendel

Please visit our website at sdhspress.com.

Printed in the United States of America

19 18 17 16 15   1 2 3 4 5

For my beloved and plucky sisters,
Clarice and Evelyn, to whom I owe so much

And for my children, Sam and George, who
make canoeing the river of life meaningful, and
Becca, who is not in this book,
except on every page

Listen natives of a dry place
from the harpist's fingers
rain

W. S. MERWIN

CONTENTS

. . . .

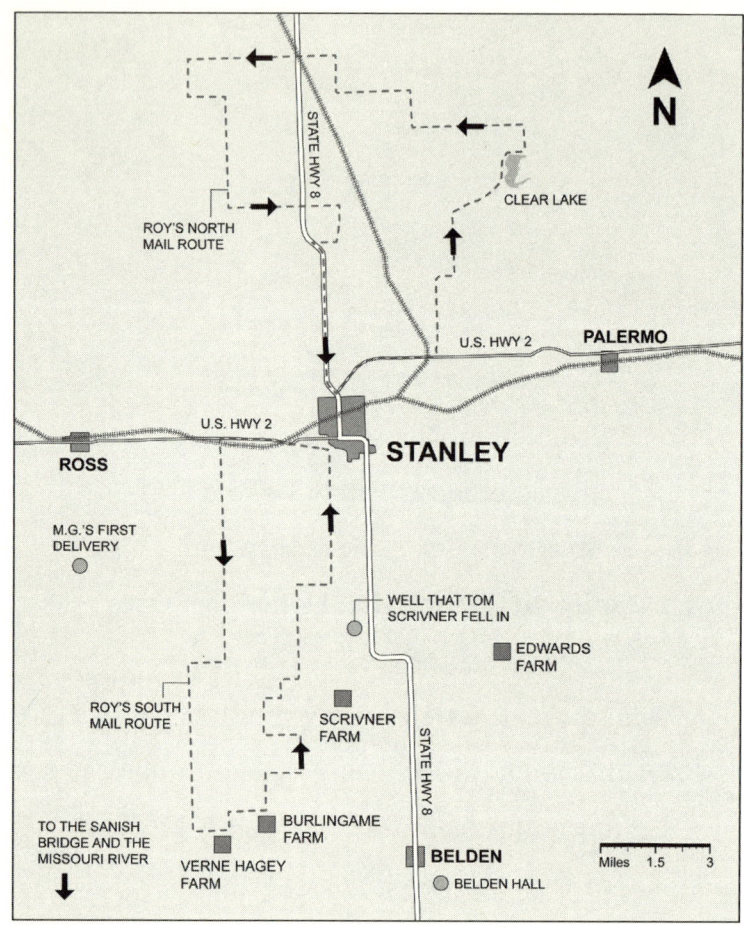

Our World, Stanley and Environs

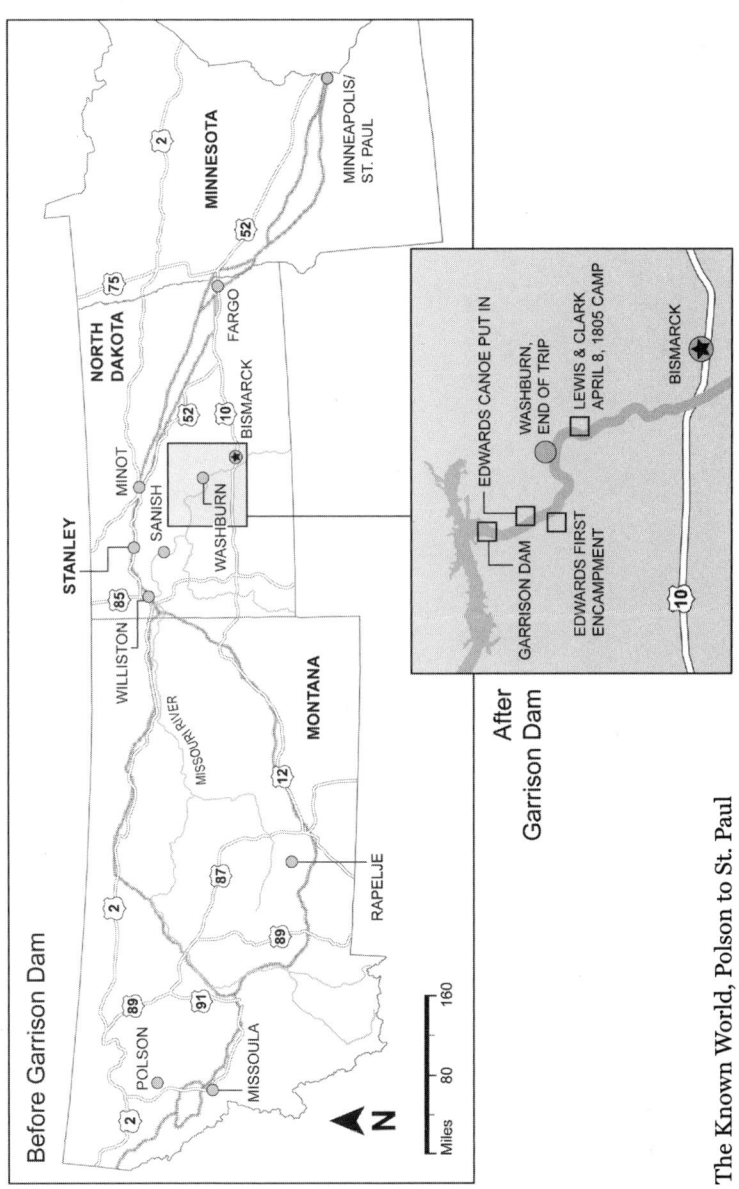

Before Garrison Dam

After
Garrison Dam

The Known World, Polson to St. Paul

ACKNOWLEDGMENTS

· · · · · · · · · ·

I conducted twenty-two in-depth interviews with surviving principals, to whom I am deeply grateful and without whose cooperation this book would not have been possible. I thank all of them for their generosity in providing information. Most crucially, I thank my sisters, Evelyn Hazen and Clarice Meacham, and my aunt, Irene Springan, for their patience and generosity in sitting for lengthy interviews and for sharing their inspiring lives with me. My brother, Jack Edwards, my sons, Sam and George Edwards, and Betty (Anderson) Bergman also generously allowed me to interview them.

My brother, Tom Edwards, answered many questions and kept me from making errors. Warren Flath greatly helped me by providing many photographs and much useful information about his family. I benefited from access to two oral interviews of Dr. M. G. Flath by the North Dakota Oral History Project and by his grandnephew Tom Flath. Verne Hagey, now deceased, provided the information for the Tom Scrivner story.

A number of colleagues read various versions of the manuscript and gave me good suggestions. George Wolf, Jonis Agee, Linda Pratt, Bill Pratt, Michael Farrell, and Wendy Katz all helped. My family provided some of my best and toughest critics, especially sister Ev, son George, and daughter Becca.

I obtained information and images from the State Historical Society of North Dakota, Betty Bergman, Warren Flath, and many family members. For help in mapmaking and handling images, I thank Katie Nieland and Michael Cooper.

At the South Dakota Historical Society Press, I want to thank director Nancy Tystad Koupal for taking a chance on this project and associate editor Rodger Hartley for excellent, detailed, and sometimes passionate editing of the manuscript. I also gratefully acknowledge support from the University of Nebraska.

Despite the best efforts of all these folks, there undoubtedly remain errors and infelicities in the book, and, lacking others to blame, I reluctantly take full responsibility for them.

This book is part memoir, part oral history.
Its chief source of information is memory, mine and
others', with the richness of memory but also its inevitable
unreliability. In every case where a clear fact has been
asserted or a quotation taken from other sources, I have
verified the fact or quotation by consulting the relevant
documentary record, for which sources are listed at the end.
But the book, and certainly its most important portions,
is overwhelmingly a work of memory, and the reader
may want to keep that in mind.

Additional information and photographs are
available at nativesofadryplace.com.

NATIVES OF A DRY PLACE

INTRODUCTION

. . . .

NEW STANLEY AND OLD

. . . .

A vast oil boom engulfs western North Dakota. Geologists had known for decades that there was oil deep underground, locked in the shale of the Bakken Formation, but it was like the sword in the stone—no one knew how to get it out. More recently engineers have devised technologies, including horizontal drilling and hydraulic fracturing, to reach Bakken oil and draw it to the surface, and high oil prices in the late 2000s finally made the endeavor worthwhile. In 2007, the United States Geological Survey (USGS) estimated that there might be three to four billion barrels of recoverable oil, putting the North Dakota field on a par with the Ekofisk discovery that made Norway a major oil exporter. In 2013, the USGS revised its estimate to seven to eight billion barrels, more like the celebrated east Texas field featured in the movie *Giant*. But some industry geologists believe that the number could be as high as twenty-four to thirty-six billion barrels, making the Bakken bigger than Prudhoe Bay and in a league with Nigeria, the world's sixth largest oil exporter. Some geologists estimate the total amount of oil in the Bakken shale, including oil not recoverable with current methods, at several hundred billion barrels, putting it on a par with the biggest world producers. Whatever figure turns out to be correct, it's big.

The Bakken strike is a boon to the American economy, pumping more than a million barrels of oil into the nation's supply each day. As recently as 2007, the United States had

1

net imports of about twelve million barrels of crude oil and petroleum products per day; in 2015, that figure dropped to about five million barrels per day, thanks in part to North Dakota oil. The United States has especially reduced its imports from OPEC member nations, and greater independence from these countries is often considered to be desirable on political as well as economic grounds. For this reason, many view the discovery and exploitation of the Bakken field as a golden opportunity, made all the luckier because the oil is located up there in unimportant, remote territory where few people live. California, by contrast, with a land area roughly twice that of North Dakota but a population more than fifty times greater, has recoverable oil reserves perhaps equal to the Bakken's, but much of it remains unexploited and off-limits due to state and federal restrictions. In the Bakken, it's full speed ahead.

The oil development has created new millionaires almost overnight. Larry Lystad, a retired science teacher with a well on his property, was reported to be expecting as much as one million dollars a year. His Scandinavian and German homesteader ancestors "have been farming rocks for generations," he said. "It's like winning the lottery." Derald Hoover, who for many years worked for the rural electric company, receives money from three wells. "It's not very hard to be a millionaire nowadays," he said. Doug Kinnoin, a farmer whose father was reported to own shares in ten wells, was asked if that makes him and his family millionaires. "I guess so," he replied.

At the center of the Bakken development are Mountrail County and its county seat, Stanley. I spent my first twelve years in Stanley when it was a tiny, dusty Dakota wheat town. My family and I left in 1956, but the town stood largely unchanged for the next half century. When I visited in 2007, the Dakota Drug was still there on Main Street, Springan's Furniture was doing a quiet Saturday business a block away, the majestic courthouse still stood amid its outriders of ash trees and a few surviving elms, and folks gathered as usual for morning

Oil tanks, trucks, service vehicles, and industrial development crowd out Stanley's "Welcome," 2014. *Richard Edwards photograph*

coffee and pancakes at Joyce's Cafe. The *Mountrail County Promoter* still published from its office on Main Street.

Driving around town, I noticed a few changes: a water park had sprung up, financed by a Stanley resident who went to Reno and made good; a new building had replaced my old grade school; Highway 2 no longer ran straight through town but swung around on a bypass. A Cenex station had sprouted on the highway, making it possible to buy gas on Sunday, and a heritage-tourism site, Flickertail Village, had been created to entice motorists to stop and spend a few bucks. After its congregation dwindled, our beautiful little Presbyterian church had been repurposed as a community space called the Sibyl Center. Mainly, though, Stanley had changed little since my boyhood. The people were now talking on cell phones and driving air-conditioned pickups, of course, but underneath they seemed much the same as I remembered, though perhaps that impression was deceptive. Even the population seemed stable. Stanley's 2010 population of 1,458 souls almost exactly matched its 1950 count of 1,486, and the feel of the town seemed familiar to me.

Visiting what I call "Old Stanley" in 2007 was like visiting

New Orleans before Hurricane Katrina or the New Jersey shore before Superstorm Sandy. Returning in 2014, I could see that the oil boom, like a typhoon, had swept away the town that I remembered. For a time, Stanley had been at the epicenter of the boom, and as oil does almost everywhere in the world, it had brought disruptive, transformative change, straining friendships and rupturing community ties. New folks, truckers, drillers, construction crews, heavy equipment mechanics, project managers, roustabouts, land and leasing agents, supply and logistics men, and all manner of other oilfield types flooded the town, overcrowding the supply of housing, filling the motel far into the future, adding kids to the schools faster than new teachers could be hired, and bringing with them industrial traffic on a scale that had been unimaginable before the boom. The pressure on housing affects the whole region; one national survey in 2014 found that Williston, an oil hub seventy miles west of Stanley, had the highest rents in the nation. A seven-hundred-square-foot apartment that would have cost $1,504 per month in New York or $1,537 in Boston cost $2,304 in Williston. Former single-family houses were carved into makeshift accommodations for the overwhelmingly male labor force. Late one weekday morning in 2011, I drove past our old house, the three-bedroom, one-bath structure that my father, Roy, had built in 1928. Now eight mud-spattered suvs and pickups filled its littered, untended yard. Roy's handiwork had become a mere dormitory.

The intense demand for housing has led drilling companies and oil-field suppliers both to convert existing buildings and to throw up new structures. One phenomenon is "man camps," temporary compounds typically built of interlocking modules trucked in from elsewhere and offering basic shelter and facilities. In the nearby Williams County town of Tioga, population 1,230, Capital Lodge and Tioga Lodge were built on opposite sides of the highway just southwest of town to house a couple of thousand workers. Stanley has two more man camps, and the dozen in Mountrail County were built to hold more than four thousand workers. Providing water and septic disposal often creates challenges, as does the mingling of transient workers. Cindy Marchello, a field coordinator for a trucking company and one of the few women climbing around the rigs, recalled that at one job, the boss didn't want to hire her because the only accommodations available were at a man camp. Marchello responded, "For every one man who's running naked up the hall, there're five men who don't want to see it either, and appreciate that there is a woman in camp so they have an excuse to tell him to put his pants back on." The boss said, "You're absolutely right. I'm one of the five. You got the job." Another woman good-naturedly complained about being a girl in the oil patch: "Port-a-potties really suck. They suck really bad!" Her female co-worker added, "Use a bush or hover."

The surge in a transient population has also meant challenges for law and order, and the upswing in crime caught the attention of the national media. The Associated Press reported, "Crime boom follows oil boom: Northern Plains towns struggle with problems that they have never had to worry about before." As one Williston official put it, "There is a bit more testosterone right now than the town was used to." Today's problems repeat on a larger scale the experience of the early 1950s, when Stanley felt the ripples of a much more modest oil strike near Tioga. I was in third grade, and we all feared Billy, a new seventh grader, a tough, mean oil kid who got into lots of fights and seemed ready to beat anyone up. If I saw Billy a block away while walking around town, I made sure to keep at least that much distance between us. I worried that someday I would turn a corner and there he'd be, and I'd have no chance to escape. Happily, he took little notice of me, and his biggest moment in town came early one evening as about ten of us younger kids trailed behind him and my friend Jimmy's big sister as they walked towards her family's combination detached garage and two-seater outhouse. "I'm going to jack Billy off," she said excitedly. None of us knew what that meant; it certainly wasn't Stanley talk, but it seemed thrilling and naughty, and it added to Billy's reputation. We were disappointed when she closed the garage door and we couldn't see. When Billy left town after about six months, his dad having moved on to a new job, it was a happy ending for us, too.

Today, the crime problem is bigger than Billy. Towns throughout the region report increases in drug trafficking, gun crimes, and assaults, and the region has seen a huge rise in applications to carry concealed weapons. Most of the newcomers are men, and most are either single or away from their families, so it's hardly surprising that prostitution and sexual violence have exploded across the landscape. Laura Gottesdiener, a freelance journalist and a newcomer to North Dakota, wrote, "When I arrived for my first shift as a cocktail waitress at Whispers, one of the two strip clubs in downtown Williston, I didn't expect a 25-year-old man to get beaten to death outside the joint." She went on to describe the toxic atmosphere of a region where, according to oilfield workers she met, "there are only two things to do in Williston: work and drink." The ramping up of booze, sex, and violence is rapidly changing the culture of towns in the region. In Stanley, more people lock their doors at night and even lock their cars on Main Street. "Before, it was unheard of to take your keys out of the car or to lock your doors," county sheriff Ken Halvorson noted. "That's starting to change, and I don't think people like it." It has proven difficult for police and emergency services to keep up with the changes. The immense rise in traffic has contributed to a 300 percent increase in ambulance calls. There are few takers for police or deputy sheriff positions, which pay less than twenty dollars per hour, while oilfield jobs pay two or three times that much.

The man camps have created such problems that despite local officials' eagerness to support the drilling, many communities imposed a temporary moratorium on opening new ones. Williston tried to stop people from living in RVs; Mayor Ward Koeser grumbled, "The people who live here and the ones who pay taxes are developing a lot of frustration with it." The demand for housing continues to intensify, however, so any limits serve only to displace the pressure to another community or another problematic fix. Companies buy single-family homes and convert them to bunkhouses, and high rents and skyrocketing house prices have placed a heavy burden on the residents who are not getting rich. It's reported that Stanley's Lutheran church lost at least six long-established families who either couldn't afford the rising costs or couldn't resist the chance to sell their houses at the inflated rates. When *Mountrail County Promoter* columnist Carol Ann Jones received her 2014 property assessment, it nearly sent her over the edge. Her house, which she describes as a "little bitty house with its 25-year-old roof, 50-by-140 foot lot on an alley that has craters large enough to swallow small cars," had been assessed at $23,000 in 2001; by 2013 her assessment had more than tripled to $69,900, and in 2014 alone its value increased an additional 65 percent to $115,300. She estimated that her 2014 taxes would be five times what she had paid in 2001 and wrote that "the City of Stanley has unilaterally declared war on property owners." But the City of Stanley isn't the problem—it's the oil typhoon.

Even though most newcomers arrive without families, the schools, too, are experiencing burdensome increases in enrollment. Recent years have seen growth of 100 or 120 students, adding 25 to 30 percent each year, and in coming years as many as 180 new students per year are anticipated. Adding to the disruption, many students enter the school system only to transfer elsewhere when their parents' jobs change. The growth requires new classrooms and teachers, but attracting teachers is difficult, partly because housing is expensive and difficult to obtain. Local districts sometimes find themselves offering housing as part of employment contracts, just as their one-room country schoolhouse forerunners often did. Taking the same cue, the Stanley Police Department attracts recruits by advertising, "affordable housing for candidate available."

Out on the prairie, drilling rigs and big trucks choke the

Tanker trucks bring oil to a collection station before
transshipment, 2014. *Richard Edwards photograph*

The new prairie landscape around Stanley, 2014.
Richard Edwards photograph

roads, and oil tanks and piping facilities sprout everywhere. I drove my dad's old south mail route past the Strobeck place and saw the familiar names from a half century ago—Bakke, Niemitalo, Piepkorn, Tiisto, Johnson, and more. In the old days, one saw golden fields of wheat swaying in the constant Dakota wind and patches of rolling prairie, still unplowed, with their exuberance of grasses. Now the peaceful landscape is interrupted by drilling rigs, new roads slashed across the fertile black prairie soil, bright flares where natural gas is burned off, tanker trucks and equipment vehicles of all types, and SUVs and pickups hauling workers to the rigs or back to the man camps. It is a busy, nearly industrial landscape. Bert Hauge, a cattle rancher, says that his family's mineral rights don't compensate for the disruption. "I enjoy the nice, quiet, out-in-the-country type of deal," he said. "You can't do that anymore."

Increasingly, the prairie suffers oil spills, saltwater discharges, and other pollution, an inevitable result of oil extraction. The state recorded nearly three hundred spills in less than two years, the oil seeping out of pipelines and tanks and poisoning the black prairie soil. The largest spill occurred in a remote area near Tioga, where an underground pipeline broke. The spill remained undiscovered by the pipeline company, but Steven Jensen, who farmed the land, found a six-inch spurt of oil bubbling up from under the ground. The broken pipe had spewed out 865,000 gallons in the two weeks before Jensen discovered the break. With only minimal requirements to report spills, oil companies rarely tell the public about them.

The risks of the oil boom are not just environmental, but civic and economic, as well. Some predict that once oil companies have finished drilling the wells and laying the pipelines, they will not need so many workers, and then the new apartment buildings and schools will stand vacant and the traffic will thin. Others are not so sure. Estimates of the amount of oil in the ground suggest that the jobs will last a long time. Moreover, the drilling method used to reach Bakken oil, hydraulic fracturing, or "fracking," is unlike conventional methods that result in years of steady, automated pumping after an initial flurry of labor-intensive drilling. Instead, the flow out of each fracked well diminishes quickly, perhaps by half in the second year, necessitating repeated drilling and re-fracking. If oil prices will support it, drilling activity in the Bakken could last for many years.

That, however, is a big *if.* In 2014, new clouds gathered over the Bakken as oil prices suddenly collapsed, plunging from more than one hundred dollars per barrel to fifty dollars per barrel or less, then settling on a plateau of between fifty and sixty dollars per barrel. National and world oil markets were awash in excess oil, flooded by new supplies from the Bakken, the Canadian tar sands, and new strikes elsewhere. Meanwhile, demand declined in response to more efficient automobiles, as well as slowing economies in Europe and China. The United States had so much excess oil, it was reported, that the country was running out of tanks to store it in. Expert prognosticators were uncertain about how long the glut might last and how far the price could fall.

The problem for Bakken producers is that their oil is relatively expensive to extract. When crude sells for one hundred dollars per barrel, Bakken oil creates great riches, but at sixty dollars per barrel, it is probably not worth further drilling. Middle Eastern countries, by contrast, have low production costs, roughly half the American level, giving them the whip hand in setting world prices. The United Arab Emirates' energy minister warned that in an oil glut, "shale oil drillers and other higher-cost producers should be the first to scale back production," suggesting the low prices are part of a strategy to limit competition from shale production. The wells that have already been drilled in the Bakken will keep producing, but fewer new wells are likely to be drilled. In 2014, there were 174 rigs in operation; a year later, companies idled so many that only 80 were still in use. Bakken producers applied for fewer drilling permits and sharply cut their offers for new oil leases.

All the big players—Continental Resources, Conoco, and others—announced layoffs and major cuts in their exploration budgets. Since it is the well-drilling activity that generates the most jobs and job seekers, the local economy is likely to suffer.

These towns remember how the mini-boom of the 1950s petered out, leaving them with unfinished infrastructure projects and debts. "We had streets and sidewalks that led to nowhere," Dickinson city commissioner Klayton Oltmanns recalled. Other communities have seen their hopes dashed in similar ways. In 1978, I visited Casper, Wyoming, to give a talk during the OPEC crisis. Oil companies were frantically drilling in the area in response to high prices, and the locals were convinced that the world's oil shortage and Casper's prosperity would be permanent. "There's only so much oil in the ground," they repeatedly told me, as though little further analysis was needed. But after peaking in 1980, oil prices plunged and remained low for the next twenty-five years.

Perhaps the Bakken will go the same way, or perhaps prices will rise again and the boom will pick up where it left off. Whatever happens, Stanley will have been forever altered. The boom has created a startling new disparity of wealth in town. Half of the royalty checks from local wells are sent out of state, many to Texas, but the money that stays in North Dakota has created a new class of millionaires. Many locals have struck it rich, amassing more wealth overnight than most Stanley folks earned in a lifetime. One local lawyer estimated that half of Stanley's residents were receiving oil checks. Yet the other half, their friends and neighbors who seem equally worthy but not so lucky, get nothing and watch in understandable dismay. "They say it'll trickle down," grumbled Jay Mackey, owner of a crop-dusting company. "The only way it'll trickle down to me is if I get a tin cup and stand on the highway."

It's as though an airplane had flown over Stanley and scattered lottery tickets across the landscape—many winning tickets but also many duds. Town residents who don't own land haven't shared in the bonanza, and even some farmers have been left out. Mineral rights are separable from land ownership, and over the years some farmers sold these rights. Now the rights-holders can come on the land and extract the oil without the landowners' permission and don't even need to pay them for it. "It's the good, the bad, and the ugly," Mountrail County extension agent James Hennessey said of the boom. "The good is the guy with seven wells who's a millionaire in twenty-four hours. The bad are those who own the land but not the minerals underneath, and whose roads are tore up and they're not getting anything. And the ugly is the disparity, which creates a lot of animosity." Mark Ellis is one of the unlucky farmers who ranches ground for which he doesn't own the mineral rights. The oil companies typically offer a few thousand dollars to such landowners to compensate for the inconvenience of drilling. For Ellis, the most aggravating part is not the dust, noise, and traffic; it is watching pumpjacks working his land, earning money for other people. "It's more interesting when you're getting a piece of it," he said.

Some winners seem pleased with their new wealth. Fred Evans is unapologetic about his new riches. A successful landowner before the boom, Evans was long convinced that the Bakken Formation would someday yield up its treasure, so he bought mineral rights throughout the area. Some complain that he took advantage of the sellers' ignorance, others that he often made his pitch just when he knew a farmer was in trouble. According to a visiting reporter, Evans's neighbors describe him as both "the richest man and the biggest crook in Mountrail County." Yet others among the newly rich feel more ambivalence about their success. Stan Wright is a former farm-equipment salesman who once allowed two farmers to pay their bills in mineral rights. Now he struggles with guilt over his windfall, insisting that he never took advantage of anyone. "It's not all good," admits Leslie Anderson, a farmer who has done well in the boom. "There are lots of families fighting that got along before." Even Fred Evans's sisters took him to court over the ownership of family mineral rights.

If Cindy Marchello and Fred Evans are the face of the New Stanley, Len Dibble, a friend of my brother Jack, was a true face of Old Stanley. Dibble's ancestors settled in Austin Township, an area that had some socialist and communist sympathies during the 1930s—which may explain his unusual first name, Lenin. Before he died in 2013 at age seventy-eight, Dibble was acknowledged to be the smoothest one on the floor at the occasional Sibyl Center dances, especially when Stan Wright's band played swing music from the 1940s. A retired farmer, Dibble lived quite contentedly in a simple trailer house on his Social Security checks and the income from leasing his farmland. Late in life he began to receive oil royalties, sometimes as much as eighty thousand dollars a month. He told a reporter that he didn't need the money, which he saved for his adult children. Nor did he need new concerns, such as locking the door to his house and remembering to remove the keys when he parked his car. And the oil boom? "I wish it had never happened," he said.

Donny Nelson, the grandson of homesteaders, farms and ranches on land south of the Missouri River. His family sold most of their mineral rights decades ago, so although he receives some small payments, they don't compensate for the disruption that the boom has brought to his land. "I'd give it all back if I could for all the trouble it's been," says Nelson, who has fought with the oil companies over drilling waste, saltwater spillage, and leakage from old storage tanks. "I don't like what it's done to our communities and lifestyle," he says. "We had a good life, and now it's gone forever, or at least for my lifetime." And Nelson says he's not alone: "Just about anybody I talk to that's a neighbor—and some of them are getting wealthy—are sick of it. . . . They're starting to realize that we had it kind of good. . . . We had a good life up here."

On a flight from Denver to Bismarck in the summer of

2013, I sat across from a robust-looking middle-aged woman who was returning to North Dakota. I asked where she lived.

"Ray. It's a small town east of Williston."

"Been there long?"

"I grew up there, graduated high school there. Except for when I went to school, I've never lived anywhere else."

I asked her how the oil development in the region was affecting her.

"It's terrible. It has ruined our lives."

"How so?"

"You can't go out at night, it's not safe, the crime, rapes, fights. We never had that before. Now we're all scared, in our own town. Never would have believed that about Ray. And the cost of things. Everything costs so much, it's crazy. Of course house prices are way high, but if you don't want to sell your house, how does that help you? It doesn't. It only raises your taxes. The ordinary people are being driven out, can't afford to stay. Or don't want to. This oil development has destroyed our lives. I wish it had never happened, but we can't go backwards now. It'll never be the same. It's just a shame." As Mary Kilen, editor of the *Mountrail County Promoter*, ruefully summed up, "The changes are big and small. They may be good or bad. Whatever happens, life as you know it will not be the same."

The Old Stanley of my memory, like many towns throughout the Great Plains, was already being effaced by death and time and by those larger forces that erode the uniqueness of small places everywhere: the loss of local civic connections because of television and the Internet, the rise of hyperindividualism, increasing exposure to mass-market culture, and a growing national culture that divides us into celebrities and nobodies in what some have called a "winner-take-all" society. In Stanley, however, oil is speeding up this process, demolishing the last vestiges of the culture I knew with a destructiveness more powerful than the grasshopper plagues of old, which it sometimes resembles.

In times like this, it is instructive to pause and reflect upon

The Mountrail County Courthouse, 1945.
Mountrail County Historical Society

what is being lost. As a child, I thought of my town—as most
children probably do—as just an ordinary place. In many
ways Stanley *was* ordinary; certainly no famous people called
it home, and no events of worldwide importance happened
there, at least until the oil discovery. Yet I have come to think
that there were exceptional things in the lives of its people and
especially in the values and virtues that they believed in and
aspired to. Just as we are sometimes blind to a painter's talent until after he or she is gone, so too we may overlook what
was extraordinary about Old Stanley and places like it until
they pass from sight. But unlike painters, who leave their art
behind to testify to their genius, the folks of Old Stanley leave
only us, their descendants, to sustain their legacy.

What did these men and women value and try to practice
in their lives? What lent their lives meaning and purpose and
gave them satisfaction and joy?

To begin with some background: Mountrail County lies in
the northwestern corner of the state, separated from Canada
and Montana by only one county on each side. It's bounded
on the south by the Missouri River. Lewis and Clark and their
Corps of Discovery paddled up this part of the river in mid-

April 1805, encountering Mandans, Arikaras, Hidatsas, and Assiniboines. Towards the end of the nineteenth century the pressure of white settlement reached this region, just as Congress passed the Dawes Act, which sought to break up Indians' communal or tribal ownership of their land by placing individual Native families on their own properties or "allotments" and then opening the remaining reservation to white settlement. Would-be homesteaders clamored for "excess" tribal land in the region to be opened to them. In response, the federal government engineered a series of land grabs. Finding that the size of the reservation was "entirely out of proportion to the number of Indians dwelling thereon," in 1891 the government took about two-thirds of the tribes' remaining land, including most of present-day Mountrail County, and made it available for homesteaders. The previously dammed-up white migration now flowed freely in a great Mountrail land rush.

Mountrail had virtually no white population in 1900, but by 1920 it recorded 12,140 residents. (The county's Indians, who had not yet been granted citizenship, were not counted.) The new settlers came from all over, the largest number from nearby states but others from the East and even abroad. Western Europe provided the largest number of immigrants, but a group of Finns established themselves around Belden, and Bohemians settled south of the tiny crossroads town of Manitou. A group of Syrians took root around the township of Ross, some coming from the eastern United States and others directly from the Middle East. Norwegian Lutherans settled the village of Palermo, which they named in honor of the Italians who built the railroad that passes through the town, though perhaps they did not consult the Italians before settling on the pronunciation *PAL-er-mo*.

Palermo, nine miles east of Stanley, is memorable to me because in 1950, when I was six and my best friend Jimmy seven, we went there for a basketball tournament. Seventh- and eighth-grade teams from all over Mountrail County would be there. Each of us tucked a dollar away in a safe pocket, and

My friend Jimmy and I both grew up as town kids in Stanley. *Richard Edwards collection*

early that day we walked over to the Stanley depot and bought our tickets. The train fare was eleven cents each way, and you rode in the caboose of the three-car train. It was a cold, gray, early spring day with a lot of snow and ice still on the ground, but we arrived without incident. We watched a couple of the early games in the Palermo school gym, then walked to Shorty Krabowski's general store, the only place in town to buy food and snacks. Actually, it was the only place in Palermo to buy anything. We splurged on hamburgers and hot chocolates and added some gum and candy besides, handing in our two dollars. The food cost us a bit more than we figured, because Shorty only returned a dime to us. How were we going to get

Bird's-eye view of Stanley in 1909.
Mountrail County Historical Society

home? We went out walking around, cold and dispirited, and had almost reached Shorty Krabowski's again when Jimmy spotted, lying there in the snow beside the sidewalk, partially covered but still completely visible, a *five-dollar bill.* He quickly stuffed it in his pocket. We were saved! When we got home, everyone was astonished that we came back with more money than we had started with.

As the different ethnic groups settled in the region, they found that life on the plains was brutal, with long, freezing winters and dry, searing summers, fields littered with rocks that had to be moved out of the way, crops often lost to drought or wind or disease, and accidents that broke limbs and strained backs, all endured in the deep isolation of the featureless prairie. For many, it was a miserable way of life. Water was a particular problem. Homesteaders who had a stream or spring on their land had a big advantage, but those without water had

to haul it, an exhausting task that engendered a mentality of shortage and worry. Soon settlers turned to wells for their water supply, but to reach water some wells needed to be hundreds of feet deep, while others, equally expensive to dig, proved to be dry. The settlers included Rosabell and Webster Edwards and Frankie and Bill Burlingame, my four grandparents. The Edwardses homesteaded five miles south of Stanley in 1902, and the Burlingames settled on a farm eight miles southeast of Stanley in 1903. My parents grew up on those farms.

Living in a semi-arid country stresses not only crops and livestock but people, too. No wonder homesteaders often fled to town once they had lived on their farms long enough—originally five years, later reduced to three—to qualify them for title to the land. My four grandparents were among them, drawn to town by the harshness of farming in that climate, the need to be closer to the high school, or simply the sheer lone-

The east side of Main Street in Stanley, 1910.
State Historical Society of North Dakota

liness and boredom of life out on the prairie. The Edwardses left first, selling their homestead in the 1910s. The Burlingames lasted twenty-two years, moving to town in 1925 after my grandmother, one of the original cleanliness freaks, finally tired after years of incessantly re-wallpapering the sod walls of their first house and then scrubbing the board floors of the wood-frame home they built to replace it.

So we were a town family. Although my grandparents had migrated to North Dakota for the free land, by the time my siblings and I were growing up less than half a century later, we had no ties to the land at all. It never seemed odd to me that we were fully town people—indeed, it gave us that slight but pleasing air of superiority that town residents almost everywhere feel toward country folks.

Stanley had been platted by one George Wilson in 1901. Like so many prairie settlements, it started as a railroad town, promoted by the Great Northern. My aunt Orba described Stanley as she saw it when she arrived as a small child in the late summer of 1902:

NATIVES OF A DRY PLACE

There was a general store very small and a pool hall where they served meals to men but of course no women never went there. There was no restaurant where women could eat. There was no post office [building], no drugstore, no bakery, no meat market, no clothing store where you could buy shoes or any clothing except a few work clothes for men. . . . Of course there was a lawyer's office and a land man who was doing a good business. There was no hotel as such. And no lumber yard. That was why [on our homestead] we had a sod house. No place to buy lumber less than 50 miles by team of horses over country with no roads. . . . There was only about 50 people in town.

Stanley was incorporated in 1908, benefiting from the homesteading boom on the surrounding prairie. At its first census in 1910, it recorded a population of 518.

Stanley secured its future early. Local folks disliked being part of Ward County because they had to travel sixty miles to Minot, the county seat, to do their official business. They campaigned to be their own county, and in 1909, Mountrail County was officially organized. This event touched off a fierce contest, as Mountrail's villages vied to be the new county seat. Local newspapers and citizens all promoted their own towns, often by pointing out the terrible flaws of the other contenders. Stanley boasted that it had over two miles of concrete sidewalks and no boardwalks, as well as more brick buildings than any other town in the county. Plaza, a competitor, charged that Stanley had such poor water that county officials temporarily stationed in Stanley had to pay the exorbitant price of two dollars a day for bottled water. On November 8, 1910, Stanley triumphed, garnering 783 votes against 517 for Palermo, 479 for Ross, and 421 for Plaza. As the county seat, Stanley would grow, while the other settlements would remain tiny villages. In 1915, the county opened its ornate and expensive French Renaissance-style courthouse just north of the railroad tracks, assuring Stanley's permanence.

LAYING THE CORNER STONE ON THE NEW MOUNTRAIL COURT HOUSE STANLEY. N.D. - KLOSS MAY 30. 1814.

Laying the cornerstone of the Mountrail County Courthouse, May 30, 1914. *Mountrail County Historical Society*

During the 1910s and 1920s, Stanley grew into the principal town of the county. The city installed streetlights and chartered a telephone exchange (our telephone number as late as the 1950s was 92); a local entrepreneur opened a movie theater; and in 1917 the girls' basketball team under Coach Isabel Flath won the state championship. In 1937, the town paved Main Street with concrete and asphalt; for several decades it remained the only paved street in town. As I was growing up in the 1940s and 1950s, Stanley's population was stable at about fourteen hundred people.

Two newspapers served Stanley, the *Stanley Sun* and the *Mountrail County Promoter*. They were filled with town news, columns reporting who had dinner at whose house on Saturday night and whose relatives were visiting from the coast, calendars of social events, weddings, funerals, and baptisms, legal and business ads, and a smattering of state political news. The more flamboyant *Sun*, allied politically with the populist Nonpartisan League, expired in 1950. The *Promoter*, more conservative and a relentlessly positive booster of all that was and is great about Mountrail County, lives on today.

Meanwhile, after the initial rush onto the land, the population in the countryside steadily diminished. Mountrail County, whose population peaked at 13,544 people in 1930, dropped to 9,418 residents in 1950 and just 6,631 in 2000, a decline that most other rural areas throughout the Great Plains shared. I witnessed this de-peopling directly. As a child, on the rare days when I got to ride with my dad on his rural mail routes, I helped distribute mail to perhaps fifty families on his south route and another fifty or so to the north. But when I revisited Stanley thirty-five years later and retraced the south route, it seemed there were only ten or a dozen mailboxes left, most belonging to vast mega-farms worked by gargantuan machinery. Two dozen farms had displaced roughly a hundred homesteads; the land was still fully employed, but most of the people had fled.

The oil boom, by its suddenness, has brought into focus changes much longer in the making. In the following chapters I try to recapture what life was like in the Old Stanley of my memory through stories about members of my family, the Edwardses, Burlingames, Springans, and Hageys, and other residents such as the Flaths, all ordinary people who learned to get on with their lives despite the harsh climate of their homeland. They lived before paved roads and four-wheel-drive vehicles, before wall-to-wall carpeting and central heating, before wheat combines had air-conditioned cabs and forty-foot grain heads, before television, cell phones, and the Internet, and, of course, before oil. They succeeded and failed as other people do, on occasion showing extraordinary courage, love, pluck, and determination.

More than that, Stanley people, like others on the Northern

Great Plains, cultivated a distinctive way of thinking about the world and how an upstanding person ought to behave in it, a set of values or character traits or habits of mind—for lack of a better word I'll just call them *virtues*—that they admired in others and strove to achieve in their own lives. I don't mean that they cultivated them consciously, as the sixteen-year-old George Washington did when he copied out 110 "Rules of Civility & Decent Behaviour in Company and Conversation" or as trendy folks today do when they hire a "life coach." Rather, the Old Stanley virtues were attributes that people implicitly understood to be right based on their society's unspoken expectations. When they acted in accordance with these virtues, they felt good about themselves, and when they failed to do so they knew they had fallen short.

There's no evidence that folks in Stanley were better at achieving their ideals than people elsewhere. They were ordinary people, neither more virtuous nor less venal than others. Like everyone, they aspired to nobility and struggled against temptation. As we will see, Old Stanley was not free from racism, possible anti-Semitism, marital infidelity, child molestation, hooliganism, promiscuity, and even murder. To list the virtues of Old Stanley is not to say that all Old Stanleyites were virtuous.

Nevertheless, a place's constellation of values and admired virtues is important. A society usually gets what it celebrates, so if, as at present in our national society, it celebrates greed it gets greed, and if it celebrates glitz and celebrity and indulgence, it gets that. Old Stanley's values produced strong, civic-minded, hard-working, and modest people who had a sense of their own self-worth within a community that they valued because they had built it themselves.

What were these virtues of Old Stanley? They included resoluteness, a kind of implacable acceptance and embracing of what needs to be done regardless of inconvenience or possible danger, and steadfastness, the virtue, established over years, of being a reliable person whom others could count on every day. Old Stanleyites also believed in devotion to community and respected those who pitched in to improve the community's life without expecting anything in return. They liked pluck, boldness in grabbing opportunity when it came along, and they expected commitment, the virtue of keeping one's word and sticking with one's choices, in doing business with each other and especially in love and marriage. So, too, they admired an insistent, dauntless optimism and believed that people made a conscious choice to find the positive aspects of any situation. They appreciated a spirit of adventure, and even those with limited means delighted in finding ways to create excitement in their lives. Finally, Old Stanley folks expected modesty, maintaining a strong dislike of self-promotion, showing off, or putting on airs.

These are the virtues that I explore in the following chapters. I illustrate each one by associating it with a particular person or people. These individuals weren't always exemplars of their virtue from the beginning; often, they had to struggle to attain it. The virtues were not givens but rather had to be striven toward over a lifetime. Nor were these folks one-dimensional; rather, they were fully rounded characters who embodied a mix of all of the virtues, and their share of flaws, as well. Old Stanleyites also admired other habits, such as fair play; reflexive honesty, the kind that needs no moment of calculation or self-prompting; humor, the spice that enlivens daily life and the universal solvent for awkward social situations; a foundational belief in the benefits of education; and an innate patriotism. Mainly, Old Stanleyites were hardy people, not so much physically robust, though some were that too, but survivors, people who typically had great personal strength and inner resources, people with deep resilience.

This constellation of virtues, values, and habits of mind was hardly unique to Stanley, and with varying emphases it would have characterized much of the Great Plains. Similar constellations probably appeared elsewhere, too—black folks might understand these values as "traditional," the ones they recall from their family roots in the rural South, and immigrants passing through Ellis Island may have fashioned some of the same values in response to very different circumstances. In Stanley, these desired virtues grew out of its particular circumstances—the values and aspirations people brought with them when they settled there, the physical hardships they faced on the dry, hostile prairie where they had to rely on each other more than perhaps elsewhere, and the newness and remoteness of the town they built, requiring them to make their own, new society.

Old Stanley's constellation of desired virtues stands in sharp contrast to the ideals others offer today. It is different from the sharp-elbowed, aggressive, look-at-me demeanor that residents of New York and New Jersey often celebrate; or the mellow and preening narcissism sometimes cultivated in California; or the martial, honor-obsessed, and intemperate bearing seemingly valued in the (white) South. Not all of the residents of those regions fit their characterizations, of course, and neither did all Old Stanleyites fit its characterization. But cultures are different in part because a society usually gets what it celebrates, and so we need to explore what virtues Old Stanley celebrated.

The approach to life that marked Old Stanley is old-fashioned and out-of-fashion now, too reticent and respectful to compete in a culture that prizes the opposite. But the customs of Old Stanley created a fairly democratic, egalitarian, and caring society—not a perfect society, to be sure, but one that fostered lives of fulfillment in a strong community. Please don't misunderstand me: I don't claim that the virtues of Old Stanley would be a model for all times or places, nor am I trying to sell nostalgia for a disappeared past. We have come too far—I myself have come too far—and eaten too much forbidden fruit to think we could be satisfied living in a New Old Stanley, any more than we could be happy if we were miraculously transported back to some imagined idyll of imperial Rome. But could we learn something by examining Old Stanley's constellation of virtues? Can we enrich our own time by studying its causes and effects? Can we perhaps find some vestiges of Old Stanley remaining in ourselves?

Out there on the Dakota prairie, my grandparents, parents, aunts and uncles, and their neighbors tried to live the virtues of Old Stanley. They tried, through example, to pass them along to their children. We of younger generations absorbed these ideas, and they remained with us long after we left Stanley. The culture of our families and of others in the North Dakota diaspora—the jokes we enjoy, the way we approach adversity, our outlook on life—continues to reflect our beginnings. So we too are marked as natives of that dry place. That worldview, even though now attenuated and mixed with other traditions bringing their own merits, continues to provide strength and meaning to my life. Perhaps, reading these stories of Old Stanley, you will find some of those virtues in your own life, as well.

The philosopher George Santayana famously said, "Those who cannot remember the past are condemned to repeat it." Perhaps we should pray to be so lucky.

RESOLUTENESS

· · · · · · · · · ·
FALL IN DAKOTA
· · · · · · · · · ·

Verne Hagey heard about it when he was halfway through his breakfast eggs. It was a Tuesday morning in late October, and when the telephone rang, Verne automatically started to count the rings. On the five-party line, three rings meant a call for the Hageys. The phone rang twice, then a third time. When it rang again, he lost interest and went back to his eggs, only dimly aware of the fifth ring. Then the phone sounded a sixth time. Six was the general ring, calling all five families to get on the line at once. Six rings meant a general announcement—or an alarm.

The Stanley operator came on and fairly screamed, "Tom Scrivner's missing!"

She meant to sound serious but couldn't conceal her excitement.

"He didn't make it home last night."

The previous morning, Tom, just turned twenty-six, had taken a team and wagon to town for supplies. Around eight o'clock that evening, the team had returned to the Scrivner farm without him. Tom's father immediately set out to find him but could find no trace of his boy. Before sunrise the next morning, he called for help. He knew that the townsfolk would turn out for the search, but he couldn't know how severely that day would test some resolute men.

Resoluteness was a quality both needed and admired in this dry, unforgiving country. Being resolute meant that when some difficult or dangerous task needed to be done, you didn't think about it, you didn't procrastinate; you just went ahead and did it. A well-digger might need to descend deep down a narrow borehole, or a mother might need to dress her four-year-old boy who had just died of whooping cough, or the town doctor might need to drive through a midnight blizzard on the prairie to reach a woman giving birth. Resoluteness is second cousin to courage, but while courage involves a self-conscious, rational decision, resoluteness is a habit of mind developed through a lifetime of experiences. Resoluteness is the chronic form of courage, so frequently practiced that it is nearly unconscious and automatic. People are said to screw up their courage, as soldiers ordered over the top at the Somme had to do; Nelson Mandela said, "I learned that courage was not the absence of fear, but the triumph over it." By contrast, a resolute person hardly notices the unpleasantness or danger of a task. It is not a matter for pros and cons, thought and decision. It is an everyday attitude of "Well, it's gotta be done, so I'd better get started doing it." Resoluteness was a powerful asset out on the prairies, a practical and self-protecting quality for people who regularly faced unpleasant, difficult, and dangerous tasks.

Verne Hagey was my mother's cousin, and I heard Tom Scrivner's story from him. Verne was twenty-three that summer of 1923, part of the big Hagey clan that had homesteaded south of Stanley just after the turn of the century. I knew Verne only late in his life, but even then he was a colorful character. One hundred percent pro-American, feisty, and humorous, he was voluble, forceful, and riveting, generous and loyal to family and friends, and sometimes abrasive to others. He had an opinion on everything.

The last time I saw Verne, in the 1970s, he had moved to California to spend winters near his daughter Janet and her family. He seemed more conscious of his colorful reputation, and I could never quite figure out how much of it was put on for our benefit. One evening, Janet served lasagna. An exasperated Verne complained to me, "I don't know why she serves that stuff when I'm here; she knows I don't like Mexican food." Verne was not a multiculturalist. He was a wonderful storyteller, and like most good storytellers, he employed an elastic understanding of the literal truth.

The Scrivners and the Hageys were neighbors in 1923, both farms lying some eight miles south of Stanley on the dry, windy Dakota prairie. Despite a quarter century of steady homesteading, this vast plain was still mostly open and untilled. Tom's father had brought his family there in 1908, when Tom was ten. Old Man Scrivner was seventy now, and he and his wife, Awrey, farmed a quarter section with one of their six daughters, their son Tom, and Tom's new wife, Anna.

When the Hageys heard the operator's news, Verne's dad started the Model T and they immediately left for town to join the search party. It was mid-morning by the time the group was properly assembled. Many of the town's men were there, including my dad, Roy.

They established that Tom had taken his team out of the livery barn in Stanley late on Monday afternoon. He purchased some boards at the lumberyard a few minutes before five and left town. No one knew what happened between then and eight o'clock, when the team came in at the Scrivner farm, late and without Tom.

It wasn't so unusual for Tom to be tardy. He often killed time walking along beside the wagon, picking up old fenceposts, planks, and whatever else would serve as firewood. Unlike the Hagey farm and others nearby that had brush and scrub trees in the coulee bottoms, the Scrivner coulees were bare, so Tom and Old Man Scrivner made a habit of picking up whatever they could find to burn when driving along with a team and wagon. When the team had come in the previous night, the wagon had been half loaded with old fenceposts, so the searchers reasoned that Tom must have left town when it was still light, or he could not have seen what to pick up.

They set off to retrace Tom's path. There were now nearly

twenty men, and the search was easy. Too easy. Tom's father had already made two trips between his farm and town looking for Tom, and there was nothing to find. The road was no more than a wagon track, and the land on either side was mostly flat—not pancake-flat like some places on the plains, but gently rolling, broken now and again by coulees and ravines. Occasionally the searchers passed wheat or barley fields, recently harvested and now lightly dusted with the first October snows, but mostly it was unbroken prairie sod.

Having searched the whole route and found nothing, by mid-afternoon the men began to focus on a well that lay about halfway between the Scrivner farm and town. Originally dug to water livestock, it was now dry and had long since been abandoned. Someone had taken a stone boat—a few planks nailed together and used to slide big rocks out of the fields— and repurposed it as a well cover to keep cattle from falling in. The searchers noticed that the cover had been moved, leaving scrape marks in the dirt, but Old Man Scrivner said, "Well, I moved them planks. I looked in there twice, and I don't believe he's in there."

The well was deep—130, maybe 140 feet. Somebody brought a light and flashed it down. The men could see something down there but couldn't tell what it was. They lowered a stone on a string down the hole; when it came back up, it was covered with stains resembling blood. Deputy Sheriff Lund sent the stone into Stanley for the county coroner, Doctor Anton Flath, to examine. They lowered another stone, this one covered with a handkerchief, and it came up stained with dark brown spots, as well. The men waited. Six o'clock came and went. Nightfall comes early at this latitude in October, and with the dark came a chill that numbed the men's fingers.

They stood aimlessly around the wellhead, the activity of searching turned to enforced and uneasy idleness. They stamped their feet and huddled in the windbreak of the cars to keep warm. Waiting also let loose the speculation that searching had bottled up. Some declared that Tom had just gone off somewhere and would soon turn up. Others believed that he'd been slugged, robbed, and dumped down the well. A few muttered more darkly, "Them damn Finlanders."

A number of Finnish families had homesteaded the area, and relations between them and other settlers were tense. The Finns tended to stick to themselves; to outsiders they seemed clannish. They didn't want their kids associating with non-Finnish kids. But as Verne remembered, the prejudice went both ways: non-Finns looked down on them, dismissing what they disliked by saying, "Well, they was Finlanders."

The center of the Finnish community was Belden Hall, sometimes called Finlander Hall, a spare wooden building two miles east of the Hagey farm in the tiny village of Belden. The dances held there on Saturday nights often became scenes of tension. Nothing irritated the young Finnish men more than seeing a Finnish girl going with an outsider, and young non-Finnish men used to ride over to Belden Hall for that very reason. The girls were not unfriendly to these advances; Verne and his friends boasted that it was easy to take a Finnish girl away from a Finn, even at a Finnish dance. And the Finnish boys didn't like it a bit.

Some thought the ethnic tension could also be traced to local political differences. Many Finns were socialists, and Belden became a center of political activity. In the late 1920s and 1930s the Communist Party found support there, as well. Its 1928 vice-presidential candidate, Benjamin Gitlow, spoke at Belden Hall, and the well-known Communist organizer and agitator Ella Reeve ("Mother") Bloor worked in the area. Although many farmers throughout Mountrail County supported the populist Nonpartisan League, some thought that Belden's politics were too radical.

There were also more serious troubles. Sometimes Verne and as many as fifteen other young toughs would ride over to Belden Hall on saddle horses just to start a fight and bust up the dance out of pure orneriness. One fight was so bad that the authorities temporarily closed the hall. On another occasion, there had been a drunken party and a Finnish man had been fatally shot. Verne's friend Billy Hill rode over to the nearest telephone and called the sheriff, but as he rode the two miles back home he kept second-guessing his decision to notify the law. He told Verne's dad, "There's just no sense in this atall. Nobody but a fool would do this. Now there's a well over there, it's purty near two hundred feet deep. It's dry. This guy who got shot isn't any good. Why couldn't we just throw him in that well and throw some rock in after him?" Calling in the authorities seemed like asking for trouble.

Memories like this one flitted uncomfortably through the minds of the searchers as they waited around the dry well in the gathering dark and cold. An alkali moon, pale and nearly full, rose around seven o'clock, giving the prairie a chilly, silvery glow.

Finally a report came back from Coroner Flath: the substance on the rock could be human blood. Now it was clear that someone had to go down the well and bring up whatever, or whoever, was at the bottom. If it was Tom, he could still be alive, despite the odds. If he was dead, he would need a proper burial, not just this hole in the ground. There was no debate about how much sense it made to risk another life for someone who was probably already dead. It just had to be done, and Ben Sather volunteered to do it. Sather earned his living digging wells. He used a rig with an auger and buckets powered by a horse that paced round and round, and he employed one man—a boy, really—Swede Edwards, who had come with him on the search. Ben was strong and used to working with wells, so it made sense for him to make the descent, and he didn't shrink from it.

Ben went back to town and returned with his ropes. Despite his experience, going down the well was no easy or safe task. Ben knew this firsthand, because when a drill bit hit a rock, someone had to go down the borehole and set an explosive charge to clear the obstruction. Then, too, lethal gas often accumulated in wells, especially abandoned wells with no air

Carol ("Swede") Edwards
as a teenager, ca. 1920.
Richard Edwards collection

circulation. To test for gas, the searchers lowered a kerosene lantern down the well and it flickered out, confirming the danger. Someone brought a pail of water, and Verne's dad dumped it down the hole. The men hoped that the water would stir up an air current, which would lift the gas up out of the well. A few minutes later, they lowered the lantern again, and this time it stayed lit.

Sather cinched the harness under his shoulders and prepared to descend. The men turned on their car headlamps so that he could check his gear. Inside the well he would be in the dark, and at twenty or thirty feet, even the circle of light that marked the surface would nearly disappear. He hung a light from his neck and tied it at his waist to keep it from smashing against the side of the well, and five or six men paid out the rope as he climbed down into the darkness.

He worked his way down slowly. He had to be careful not to knock any rocks loose from the wall. Assuming that Tom Scrivner was down there, and hoping he was still alive, dropping a stone a hundred feet onto his head was unlikely to improve his condition. Sather's progress, measured by the rope being paid out above him, was slow but steady. In fifteen minutes he had descended seventy feet.

Then he stopped. Far short of his goal, he yelled to the men at the surface that he could go no further. The first seventy feet had been dug on a two-foot bore, but then the well driller must have switched to a smaller auger, for here the well narrowed to eighteen inches. Sather was a big man, not fat but tall and burly, and his shoulders were just too wide for him to continue down the narrower hole. He tried to maneuver to see below him, still unsure whether Tom was there. But in the cramped conditions he could see nothing. He yanked on the rope twice, and the men aboveground began pulling him up.

Coming up was faster than going down, and in a few minutes Ben Sather was standing among the search party again. It was very dark now—the stars blazed overhead and the pale moon added its weak light, but outside the small circle of headlights was the awful, lonely blackness of a plains night unrelieved by electric lights. At least the stars promised that the weather would remain clear. But the wind sprang up stronger, and the cold stung the men under their coats and gloves. In the deepening gloom, they huddled together as much to overcome their isolation as for warmth. They were reluctant to call off the search, but they couldn't see what more could be done. Old Man Scrivner had grown dispirited as the hours wore on—first the fruitless afternoon of searching, and now the helpless idleness of standing around the well. He wanted to go down himself, but he was nearly as broad as Ben Sather was, and Sather was certain that he would have no better luck getting to the bottom.

Then Ben's young helper, Swede Edwards, volunteered to go down. Swede was a regular beanpole, tall with slim hips, narrow shoulders, and no fat anywhere. Like Ben, he was strong from farm work and now well digging, but he was sinewy where Ben was burly. His real name was Carol, and he wasn't Swedish at all, in contrast to several men in the search party, who were real Swedes. But he was tall and blonde, and that's what everybody called him, Swede.

He was my dad's little brother. There had been six kids in the family, all spaced about two years apart: two boys, Edgar and Harold, then a girl, Orba, then my dad, Roy, then another sister, Birda, and finally Carol. As often happens, the older siblings had special feelings about the baby of the family, though they were perhaps not unmixed feelings.

Swede had turned twenty-one in the summer of 1923. Of course, I remember him from later in his life, and he seemed a remote and sophisticated figure. He lived in Chicago, a fabled and faraway place. He had divorced his first wife, which was scandalous by Stanley's standards, and he worked as a salesman or executive for Automatic Electric, a manufacturer of dial telephones. And although I was not yet much aware of money—of which families had it and which ones did not—I had a sense even then that Uncle Swede was rich. The story was that he had invested some money during the 1930s, when stock values were depressed, and that by the 1950s, he had become rich.

I don't know what Roy really thought of his little brother; as adults their lives moved in such different directions. They had grown up together in the same remote place, sharing their early lives and experiences, but their relationship had grown complicated. Their oldest brother, Edgar, had drowned when he was eighteen. The next oldest boy, Harold, had nearly been killed when a train hit his car, and it was years before he recovered. When my grandparents' marriage disintegrated and Webster moved to his isolated farm north of Minot, Roy became the oldest responsible male, so he dropped out of school after tenth grade and went to work. In contrast, Uncle Swede left for Chicago, where he became rich. I'm sure that Roy had a certain pride in his younger brother, but also the lingering resentment that the responsible son who stays home and makes sacrifices may feel toward the prodigal son who leaves to reap the world's rewards.

Even though Roy was usually a forgiving person, there was a period of twenty-three years during which the brothers never spoke. I never knew why and could only guess. Whether a peace was finally brokered, or whether time simply healed the wounds on its own, in the early 1950s Swede and his daughter Carol Ann came to visit my family. Carol Ann, then about twelve or thirteen, liked Stanley so much that she stayed after Swede went back to Chicago, living with us for the whole summer. She was several years older than I and just a year or two younger than my brother Tom, so she was more his friend than mine, which made me jealous. Carol Ann was pretty, and she carried the mystique of being a big-city girl. Every evening after dinner, about ten of us played kick-the-can and hide-and-seek under the single light bulb fixed on the old garage out back. Sometimes Carol Ann would hide with me behind the bushes lining the garden, which made me feel grown up. When summer ended and it was time for her to go back to Chicago, she cried and begged Uncle Swede to let her stay. But he said no, she had to return, and Stanley was less fun after her departure.

The last we heard of Uncle Swede before his death was in 1953. My oldest brother, Jack, was studying at Macalester College. Jack never had any money beyond what he earned himself during the summer, and most of that had to go toward his tuition. He would develop a lifelong habit of being both personally abstemious and perfectly contented, but paying for college was always a challenge. Uncle Swede sent Jack a letter containing a check for one thousand dollars—as much money as Jack could save from working two whole summers, a colossal sum. Jack worried about how he would pay it back, but Uncle Swede said, "You don't have to. Sometime when you're in a position to do so, make a gift to someone else who needs it." It was a generous act indeed, and perhaps also Swede's way of making restitution for his difficult relationship with Roy.

All that lay ahead of Swede. Now, as a twenty-one-year-old, in the biting October night, he prepared to go down the well and look for Tom Scrivner. He was part of a culture that expected men (and women, too) to be resolute, so that when something needed to be done, they just went ahead and did it, regardless of its danger or difficulty and without calculating the risks or balancing the probabilities. Swede wriggled into the rope harness in the dim headlights of the parked cars and then strapped on his light. He proposed to descend into a hole 130 or 140 feet deep, the last half of which was too narrow for Ben Sather's shoulders, to find there—who knew what? Possibly Tom Scrivner's crumpled, bloody body, possibly a wounded animal, possibly something else. No one knew the condition of the well, the chances of a cave-in, how much gas was still down there, or even if there was water at the bottom. It didn't matter. Like Ben, Swede had learned to be resolute; only the task mattered.

He climbed into the well, and like Sather, he made good but slow progress. After about fifteen minutes, he reached the ledge where the hole narrowed and started down. The smaller hole left virtually no space to move, but he managed to squeeze and slide his body down. His progress was much slower; he did best when he kept his arms extended straight above his head. After another forty-five minutes, he reached the bottom.

Or not quite the bottom. Tom Scrivner's body was wedged there, upright, so that Swede had to be careful not to stand on him. He was dead.

Swede yelled to the men up top that Tom was down there, and they lowered a second rope. He worked it past his own body, then began the grisly job of getting it around Tom. His

Mountrail County Promoter

XVIII, NUMBER 4. · STANLEY, MOUNTRAIL COUNTY, NORTH DAKOTA, FRIDAY, OCTOBER 26, 1923. · Subscription Price $2.00 Per Year

Horrible Death Claims Life Of Young Man Monday

...T HIGH SCHOOL STUDENT ...ILLED NEAR PALERMO SUNDAY

OSCAR O. ODEGAARD

Thom Scrivner Found Dead In An Old Abandoned Well On The Bates Homestead

B. B. HAUGAN WILL SPEAK IN STANLEY SATURDAY EVE. OCT. 27

America's Best Judge of Hogs

NORDMANDEN'S REPRESENTATIVE IN STANLEY

STATE WHEAT MEN WILL GO TO DENVER

The *Mountrail County Promoter* reports the death
of Tom Scrivner, October 26, 1923.

lamp made more shadow than light, and he used his feet more than his hands. The constricted hole left no room to work and imposed a gruesome intimacy on the two figures.

After several unsuccessful tries, Swede managed to poke a loop of rope under Scrivner's shoulders. He yanked twice on his own rope, and the men aboveground began pulling him out. First Swede emerged. Then Tom Scrivner's body, stiff, broken, and battered from the fall, was drawn from the well into the cold prairie night. The search party, relieved at having finished its task but now burdened with grief, returned to town.

Tom's death caused a sensation, which area newspapers reported breathlessly in stories riddled with error. The Minot papers—the *Daily News* and the *Northwest Press*—insisted on calling the deceased "Scribner" and claimed he was twenty-three years old. The *Van Hook Reporter* called him "Scrivener" and said that the well was only eighteen inches wide and eighty feet deep. The *Stanley Sun* said that it was Roy, and not Swede, who had gone down the well. Only the *Mountrail County Promoter* seemed to get its facts right, but it left the controversial parts of the story untold.

The town's attention had immediately turned to how and why Tom Scrivner had died, and a three-man coroner's jury was convened on Wednesday. Coroner Flath examined the body before the inquest and found that Scrivner's hands, face, and neck were scratched and bruised, but that "none of the injuries [nor] all of them together were sufficient to cause death." Tom could have been bruised during the fall, and lacking any other explanation, the jury decided that the gas must have killed him. It appeared that he survived the fall, because, in the *Sun's* words, "An arm band had been pulled from the sleeve of his right arm and was gripped in his hand when the body was recovered." Tom must have spent some hours in panic and despair, wedged in the well as though in a straight-jacket and barely able to move, in total darkness, knowing he was going to die and able to do nothing to save himself.

The coroner also discovered that Tom had twenty-five dollars and fifty-five cents in his pockets. This finding was significant, because, as the *Daily News* put it, it dispelled "a belief that had been prevalent that Scribner had been slugged and robbed and then thrown into the well." The *Sun* echoed this conclusion: "This fact alone seems to effectually dispose of the foul play rumor that was hinted while the search was in progress."

The accident was understandable, according to the *Promoter*, because the well was abandoned and "most of the people residing in the neighborhood were not aware of the well which had been dug about seventeen years ago and as there is no one residing on the place it had remained covered up with pieces of lumber." The *Daily News* and the *Van Hook Reporter* agreed that the well's existence was not generally known, so it made sense to them that Tom might have been ignorant of what was under the stone boat. The coroner's jury found that "the death of Thomas Scrivner was accidentally caused by his falling into an abandoned well on the farm of R. Baks and being overcome by gas." The state's attorney also declared the death to be an accident, pure and simple.

But the rumor of foul play was harder to dispel. Some argued that if it had been an accident, Tom would have fallen into the well head first. The state's attorney, however, theorized that Tom could have fallen in feet first as he tried to pick up the stone boat for use as firewood. He would have grasped one edge of the boat and tipped it up and then taken a step forward to prop it against his chest and grip the outside edges of the planks. In doing so, he must have stepped right into the well, with the cover slamming down over him and covering the well up again. This point was crucial, because Old Man Scrivner insisted that the planks were lying unmoved over the well when he first examined it. With such a simple and straightforward explanation, the state's attorney insisted that there was no need for further criminal investigation.

The trouble was, Verne, his dad, Ben Sather, Old Man Scrivner, and others didn't believe a word of it. Contrary to what the newspapers reported, everybody, including Old Man Scrivner and Tom himself, knew about that old dry well, and Tom was unlikely to be careless around it. He would have dragged the boat away from the well before trying to pick it up. More to the point, since he knew what was underneath, Tom would not have picked up the boat in the first place. Nobody was foolish enough to take the cover away from a deep well, since livestock grazed nearby and could have fallen in.

On the face of things, Tom Scrivner had no enemies. He hadn't taken part in any of the fights with the Finns; in fact, he never even went to the dances in Belden Hall. He was quiet, frugal, and hard-working. He was no troublemaker, and he seemed to have given offense to no one. The *Promoter*, as usual avoiding controversy and so perhaps intentionally vague on details, sustained this view and reported that "the deceased leaves his wife to whom he was married about two years ago and a child about one year old."

Underneath the surface, however, the facts were somewhat different, and Tom's friends knew about the bad blood in that neighborhood south of town. The previous year Tom had married one of the Niemi girls, Anna, who was then just seventeen. The Niemis were Finns who lived to the east of the Hageys, and they were part a big extended family. Anna's widowed mother had made no objection to the marriage, but it had not gone down well among the young Finnish men.

The deeper tension between these prairie Montagues and Capulets on account of Tom Scrivner's marriage is not hard to guess. The *Minot Daily News*, perhaps feeling less need than the *Promoter* not to tell the full truth, used liturgical cadences to cloak its earthier revelation; its November 1, 1923, obituary noted, "Deceased was married to Miss Anna Niemi July 1st, 1922, and of that union a girl baby, now eight months old, was born." Finnish roughnecks could count the months like anyone else. The baby girl born in February 1923 had been conceived back in May 1922, when Anna was just sixteen, and had been legitimated by a July 1922 wedding. Now, more than a year later, an October plunge down a well had left her fatherless.

Not too many people believed the state's attorney's theory, but no one knew for sure whether Tom Scrivner had accidentally fallen down that well or had been thrown down it by young Finnish toughs in an act of revenge. They never did know.

What they did know was that the men called out to undertake the search would not shrink from their work, nor would the men—a man and a boy, really—who quickly saw that they needed to descend into the well. There is no indication in either the written or oral record that anyone held back or flinched when it became clear what needed to be done. Perhaps Swede's act could be dismissed as just the confident imprudence of youth, but Ben Sather was a middle-aged man with a family and a business, and he, too, unhesitatingly dropped into the well, going as deep as it was physically possible to go. Both of them just went ahead and did it. Resoluteness was expected; they expected it of themselves.

STEADFASTNESS

· · · · · · · · · · ·

A NECESSARY MAN

· · · · · · · · · · ·

This day was special, because I got to ride along with my dad, Roy, on his mail route. Such a thing didn't happen often, and almost never on a school day. But two days earlier, Dr. Flath had taken out my tonsils, and I wasn't yet allowed to return to school. The day was perfect, the endless snow-covered fields stretching away clean and white, glistening in the warm spring sun. We rode in my dad's blue Ford sedan. The post office required him to provide his own vehicle, and he always bought Fords.

As usual, Roy had gone to the post office early that morning to sort the mail, preparing it for delivery to about fifty farm families on the route south of town and another fifty families on the north route. It was hard for me to get up as early as he did, so at 7:30, when he was ready to leave town, he swung by our house again and picked me up. Packages and boxes were piled on the back seat, and the mail sat between us in the front, tightly banded with a leather strap. Each time we approached a farm along the gravel road, Roy reached over and took out the next batch of mail. He pulled up alongside the mailbox, opened its door, and removed any outgoing mail, checking to see that it was properly stamped and addressed. Then he lowered the little red metal flag, put the incoming mail in the box, closed its door, and drove on to the next farm.

Roy periodically tightened the strap to keep the bundle together, but by the time we approached Dewey Jarmin's place, the last stop on the south route, there wasn't much mail left to deliver and the bundle had come a little undone. A postcard lay loose by itself on the seat, and I picked it up and looked at the exotic picture—palm trees, a deep blue ocean, and two dancers in grass skirts with their arms outstretched. I asked, "Dad, what is this?" He turned his head, saw what I was holding, and grabbed it out of my hand. "IT'S OTHER PEOPLE'S MAIL, THAT'S WHAT IT IS!" For him, the mail was a sacred trust, and we weren't meant to be looking at other people's mail, even the picture side of a postcard.

My father came to be a quietly inspirational figure in Stanley after a hardscrabble childhood and a few years as an adventurous young man. He had always been reflexively honest and possessed the essential modesty that fit his small town, but it had taken time for him to grow into the responsible family man and steadfast contributor to community life that his Old Stanley contemporaries came to admire.

The virtue of steadfastness can only be developed and appreciated over time, and only with time can you learn which people can be counted on. Counted on for what? Counted on to do what needs to be done, counted on to be there when needed, counted on to do their jobs or play their parts even when doing their jobs or playing their parts is not what they want to do, or is not especially in their own interest, or is not pleasant or easy or convenient. In a small town, as in most organizations or groups, there are people you can count on and there are other people. The other people participate when they are interested, when it is convenient, or when they have something to gain by it, and sometimes they make important contributions. Many of us from time to time are moved by the circumstance of the moment to make self-sacrifices, whether it be helping neighbors after a severe storm or responding to a moving appeal on television. But the steadfastness admired in Old Stanley was not so much a conscious, considered act as a habit, a constant appreciation of the obligations one had to one's neighbors and community—obligations that did not require calculation. Stanley had a number of steadfast people. One was W. R. Stewart, the long-serving superintendent of schools, whom the town counted on to be a leader in education and for whom they named the elementary school. Another was M. G. Flath, the town doctor. Roy was one of these people, too.

Roy was born in 1898 and grew up in a sod house five miles south of Stanley, where his parents, Rosabell and Webster Edwards, homesteaded in 1902. Their nearest neighbors were miles away. The long winters meant extreme isolation for the parents and six children cooped up in the ten-by-fourteen-foot house, and the hot summers were equally uncomfortable, though at least the children could play outside. One summer day, when Roy was nine and his little brother Swede was five, the two boys were given the chore of walking the family's cows to a nearby creek, watering them, and then driving them home again. The trip would take all day. The boys took along some matches, and as they walked home, they lit little fires in the grass and stamped them out. Fires are always fun for small boys, but to play with fire on the prairie must have been an especially exciting risk. Once a prairie fire got underway, it was nearly impossible to bring under control; only a major barrier like a river or a railroad embankment could halt its progress. Everyone, even small boys, knew to be scared of prairie fires. The boys lit one fire and stamped it out, then they lit another and stamped it out. Their third fire they could not stamp out; it burned beyond their control and spread out onto the prairie. How far it burned they never knew.

The two boys must have been terrified at what they had done and swore never to tell anyone about it. Perhaps they didn't even mention it to each other. Roy kept his secret until he was well into his sixties, when he unexpectedly revealed the story to several family members at my sister's home in suburban Virginia. Even though the fire had happened sixty years earlier, the memory of it still visibly distressed him.

The Edwards children went to a local country school that

operated only during the summer. Seeing her children growing up undereducated, Rosabell moved them to town in 1905. Even then, the children missed a lot of school because they were needed on the farm in the spring and fall. Rosabell's move also began the dissolution of the family, a slow process that would eventually result in Webster's complete disengagement from his wife and their children. By 1912, Webster had moved by himself to a rented farm north of Minot, some sixty miles away, and later Rosabell moved to Oregon, taking only Swede, the youngest child, with her. None of the children, especially Roy, ever wanted to admit the shameful fact that their parents had divorced. Few thought of Rosabell as a nice person, and Webster was mostly a recluse. The wonder was, as Roy's mother-in-law, Frankie, used to marvel, "How could someone as nice as Roy come from a family like that?"

By the time Roy finished tenth grade, his oldest brother had drowned in the Great Northern reservoir, and the next oldest had been badly injured when a Great Northern train hit his car. At fourteen, Roy was left as the family's oldest responsible male. He dropped out of school, and for the next few years, he worked to support himself and his sisters, serving for a time as a short-order cook in a small restaurant. As his sisters grew older, however, he was gradually freed of family obligations, and he began to enjoy life. He was good-looking and athletic, with a kidding sense of humor and a spirit of adventure. He occasionally played for the town baseball team, though he was limited by his inability to hit the curve, a shortcoming he passed on to his "good glove, no hit" sons. He learned to golf. He bought a car, a Model T roadster, the first of thirty-three Fords he purchased over his lifetime. In 1918 he enlisted in the army at Minot and was sent to Norfolk, Virginia, for basic training and to learn how to be an army cook. Then it was on to France, arriving too late for combat but not too late for an experience. The boy from the sod house marveled at French life, but it did not leave him envious; he only became more appreciative of the variety of human life. When he mustered out at Camp Dodge, Iowa, on February 19, 1919, he was just twenty-one years old.

Now discharged, Roy was at loose ends. One summer he joined a traveling circus that moved among small towns in North Dakota and Minnesota, setting up in each town for a few days. For his attraction, he created a canvas-walled booth featuring a six-legged calf and a dodging monkey. The calf had been born a freak and did indeed have six legs; Roy charged two pennies to see it. The monkey was set up at the back wall of the booth, and for a nickel, Roy offered passersby three balls to throw at it. If they hit the monkey, they won a prize. At one stop in a small Minnesota town, some ladies from the local humane society wanted to shut Roy's booth down on the grounds that striking a monkey with a ball was animal cruelty. In response, Roy offered the ladies (or their husbands) as many free throws as they wanted. As they soon discovered, it was impossible to hit the monkey, and they withdrew their complaint.

Around Stanley, Roy was seen as a highly desirable young bachelor, well-liked, clever, and adventurous. He was popular at the dances. When he was twenty-five, he began to focus his attentions on sixteen-year-old Winnie Burlingame, who had just graduated from high school and lived with her family on a homestead several miles outside of town. Her parents, Frankie and Bill, had no objection to this older man who had been all the way to France, but the situation changed in the summer of 1924, when Winnie discovered that she was pregnant. Bill liked Roy so much that he seemed to accept what had happened, but Frankie was outraged. "How could you do such a thing?" she demanded. Still, everyone had to put the best face on it, and Roy and Winnie rushed off to Montana to be married. Their daughter, Clarice, was born the following February. Roy's status suddenly changed from carefree young bachelor to husband and father. Like Winnie, he was intensely proud

Roy and Winnie in 1922, two years before their marriage.
Richard Edwards collection

of their new baby and traveled through deep snowdrifts with a team and wagon to show off the two-week-old Clarice to the Burlingames.

But Roy had trouble learning to be a husband. He and Winnie moved into a tiny apartment above Shulkin's Confectioners on Main Street, and Roy's sister Bird and her two children came for two weeks to care for Winnie during her lying-in.

Winnie was grateful, because Roy himself wasn't around much. She cried constantly over his absence and, being so young, found it hard to deal with a colicky baby alone. Their second daughter, Evelyn, was born nineteen months later, giving Winnie two small babies to care for. Roy remained distant, golfing or playing cards and staying out with his friends when he wasn't working. He played poker weekly with G. O. Flath, Perry Culvert, and others. Roy's frequent absence worked a terrible hardship on Winnie, who resented his reluctance to accept his new responsibilities. She rarely had any money of her own and didn't know how much Roy had. (Years later, when Rosabell was visiting, she insisted to Roy that he didn't really have to give any money to Winnie, telling him, "It's your money, you earned it!" By such pronouncements she cemented her unpopularity in the family.) For his part, Roy never said that he regretted giving up his freedom for the obligations of family, but his actions spoke for him.

After a couple of years, Roy came home one day and announced to Winnie that they were moving. This decision was a complete surprise to her, and she asked, "Where to?" He had arranged for them to move into a shabby, unpainted little house about three blocks off Main Street that had once belonged to his parents. It had only three or four rooms and no running water. It was hot in the summer and cold and drafty in the winter. And Winnie was pregnant again, with Billy.

In 1928, however, Roy took a step towards the steadfastness that would mark his later life when he decided to build a new house. It came to be called "the big house," although that term must be understood in a relative sense. He borrowed some contractors' books and, working with a friend, did everything from pouring the footings to putting in the plumbing and wiring to nailing down the shingles. When they moved in, it seemed like a dream home to Winnie, with its well laid-out kitchen, basement laundry area for her wringer washing machine, a large bedroom on the first floor, and two small bed-

Winnie at her sink in "the big house," 1942.
Richard Edwards collection

rooms on the second. Not much later Roy converted a first-floor closet into an indoor bathroom. They—we—lived there for twenty-eight years.

Slowly Roy grew into his new clothes as father, husband, and contributor to the community. From the beginning, he had relished playing with his kids, and he took delight in doing things to please them. Any new clothes that Clarice and Evelyn wore were likely to be home-sewn, but when Clarice had a special occasion at school at age thirteen, she received the treat of a new store-bought dress. Clarice went uptown and found three dresses to bring home "on approval," from which she would select the one to keep. At home, she modeled each one for Roy, who then astonished her by saying, "Hmm . . . looks to me like we better keep all three!" His comment was a huge boost to the young girl's confidence, and she did keep them all.

One Christmas, Roy arranged a special gift for the girls—a

bike for them to share. Gifts tended to be small at the time, and they were thrilled. Another time he woke the girls up early in the morning, telling them, "You'll never believe what I just found in the barn!" Eyes wide open, they demanded, "What?" He replied, "A new little calf." They shrieked as they ran to see it. Its arrival could not have been news to him, but he delighted in making everything seem filled with surprise and fun. Later he designed rubber guns for his sons, slingshot-type gizmos that shot a rubber band with some zip. The rubbers were inch-wide strips cut from an inner tube. The synthetic rubber that largely replaced real rubber during World War II didn't work, so we searched diligently for older inner tubes. While the projectiles couldn't injure you, they did make a pleasing *thump* on your victim. We organized wars with foxholes dotting the garden and paper bags filled with sand as our grenades, and, wearing Roy's old World War I helmet, we slew Germans for hours on end.

The 1930s were hard years. For most of the country it was the Great Depression, but Dakotans knew the period as "them dry years," when farmers saw their crops wither in the fields and there were no jobs to be had. One day Roy and Clarice were sitting in his car on Main Street when a woman named Mrs. Haas, whom Roy knew only slightly, came walking by. She was crying. He asked her what was wrong. "I can't go home," she said. "I just can't. My children are hungry, and I have no money. There's nothing I can do." Roy jumped out of the car and ran into the grocery store, emerging a couple of minutes later with a big bag of groceries for her.

Roy was fortunate to be able to help others, because he had a job that both provided for his family and contributed to his growing steadfastness. It was a job that had existed only for about as long as Roy himself, but it was an important one. For most of the nineteenth century, farm families had had to go to a country store or to town to pick up their mail. This errand might be combined with a weekly trip to pick up supplies, but it often imposed a real hardship, especially when it meant oming in from a remote farm over primitive roads in bad weather, perhaps to find that there was not even any mail to pick up. In 1896, the United States Postal Service launched its Rural Free Delivery service, and mail carriers began to bring the mail directly to farm families. In April 1923, the Stanley Post Office hired Roy to deliver its rural mail, and the responsibility would give shape to the remainder of his life.

He was, of course, simply a mailman. Yet it is hard now to understand how much farmers and their families depended on their mail carrier. Most farms in the first half of the twentieth century had neither telephones nor electricity. Despite a flurry of postwar rural electrification, by 1950 only two-thirds of Mountrail County's 1,484 farms had electricity and about 40 percent had telephones. When the power line arrived, it typically meant having a few lights and perhaps a radio and a couple of domestic appliances. These were prized additions, to be sure, but they hardly hinted at the transformative force that electricity would become later with the advent of television and the Internet. As for the party-line telephone, it was available in emergencies and was occasionally used for arranging social events, but otherwise it mostly sat idle. After all, Evelyn Wakefield, the switchboard operator, might be listening in, and people knew which callers stayed on the line too long gossiping. Few people conducted business on the telephone. It just wasn't the custom, and besides, all your party-line neighbors would know your business.

So as late as the 1950s, Mountrail's isolated farmers relied on the U.S. Mail as their crucial link to the outside world. It brought news from relatives, boxes of chicks to be raised, Montgomery Ward and Sears Roebuck catalogs, occasional fresh fruit, newspapers, packages with clothes or tools ordered from Chicago, farming and hunting magazines, Christmas presents, reports on crop prices, letters of legal and banking business, U.S. Department of Agriculture reports, seeds for vegetable gardens—in short, nearly anything from the world beyond the farm. In turn, the mail carried away equally important items including letters to relatives, mail orders, indeed all the connections that a family clung to. Sometimes the mail brought items essential for the farm's survival; in other cases it simply provided welcome relief from the drudgery and isolation of farm life. It was much anticipated—who knew what today's mail might bring?—and hugely appreciated.

The rural mail carrier was the critical link in this system. Today the mail carrier is regarded as just another lowly service worker, and possibly not even the most valued in the hierarchy of UPS and FedEx drivers, cable guys, Verizon salespeople, newspaper carriers, and others who keep us connected to the world. Like them, the mailman has become anonymous, and the U.S. Mail itself is derogated as "snail mail." But in Roy's time, the rural mail carrier was the vital connection to the world for the farmers he served. If he got sick, there was no mail. If he decided that the weather was too threatening or the snow too deep, there was no mail. If he were careless or tardy, half the chicks might be dead or the fresh fruit spoiled by the time they were delivered. If he were disorganized, the mortgage statement might be delivered to the wrong farm. So much depended on him.

Roy's mail routes took him both north and south of town. He started in Stanley, driving out of town on paved U.S. Highway 2. Soon, however, he turned onto the section-line roads, gravel roads or just tracks that went straight north-south or east-west along the checkerboard squares of fields. Some of these roads were maintained, a road-grader having created sloping shoulders and a shallow ditch on each side, while others were mere trails with two wheel tracks and grass growing up the middle. By the 1950s, most prairie tracks had finally been turned into dirt roads that were occasionally graded, allowing Roy to cover more ground in less time.

A farmer or his wife could see Roy coming by the cloud of dust trailing him in the summer or by the dark outline of his car on the white snow in the winter and frequently would meet him at the mailbox to exchange a few words. Roy's arrival was so regular that farmers said they could set their clocks by him; if he was a few minutes late, they would ask him where he had run into trouble. Roy's car might be the only one to pass by the farm that day or even that week, so his arrival was an event. Farm families welcomed him because he was perpetually upbeat and amusing and happy to spend two or three minutes chatting. As a farmer leaned against his car, Roy relayed news from town and they visited about the weather, what the roads were like, how the crops were doing, anything that broke up the monotony and loneliness of farm life. A few people liked to linger for longer conversations, forcing Roy to say three or four awkward goodbyes before he could drive away.

In these small, tight farming communities, privacy was much prized, and the rural mail carrier was privy to many secrets. What magazines are the Arndts reading? What did that fat letter to Matt Niematelo from a Chicago bank mean? Oddie Tiisto keeps sending money orders to his brother in Oregon; did he lose his job out there? Doesn't the letter from Senator Langer to the Niemis prove that they voted for him? Franz Wilkie still subscribes to a German-language newspaper; is he a true American? The mail carrier knew all this and more. Much as farmers or their wives wanted to chat with Roy at the mailbox and hear gossip about others ("Why does the Strobeck wheat look so poorly this year?"), they depended on him to keep their *own* affairs private. If the rural mail carrier talked loosely with other folks on the route or with people in town, everyone would know their business.

Roy understood the importance of his role and treated it as a trust, not just a job. It meant being friendly but discreet. It also meant showing up for work every day and getting the mail through regardless of the weather or the condition of the roads, and between 1923 and 1940, outside of his regular vacation, he missed only ten days of work. Finally, it meant not allowing me or himself to read so much as a postcard of Dewey Jarmin's or anyone else's mail. To be a good rural mail carrier

Roy in his Model A Ford with tractor tires on the rear wheels with his son Jackie, 1936. *Richard Edwards collection*

required steadfastness above all else, especially when circumstances made the task difficult.

Winter was the most challenging time to deliver the mail. It was not uncommon for the daily high temperature to peak below zero for many days running. One day Roy stopped at home between routes and cheerfully told Winnie, "You'd better wrap the kids up good today, it's forty below out there." The wind posed another hazard; even a modest snowfall, say six inches, could build up into drifts several feet high when heavy winds blew. The roads would be plowed out eventually, but the mail could not wait.

Roy devised several ways to deal with these conditions. At first he delivered the mail from a tiny cabin mounted on a horse-drawn sled. Inside the cabin was a small coal stove with the stovepipe stuck out the side. Later he invented what must have been one of the first snowmobiles. Four-wheel-drive vehicles were not widely available, so Roy fitted out his Model A

Ford with tractor tires in the rear to break through the drifts. He experimented with variations on this theme, one winter replacing the front tires with ski runners, and another time switching out the tractor tires for caterpillar treads. Passengers were always surprised to see how fast he drove when he hit a snowdrift, but speed was crucial. If he drove more slowly, he risked getting stuck halfway through and spending the next couple of hours shoveling his way out. All of Roy's determination and ingenuity notwithstanding, there were a few farms that simply remained unreachable for six or eight weeks during the worst part of winter, and Roy would engage someone with an airplane to fly over them every couple of weeks and drop their mail.

Together the north and south routes covered sixty-four miles. Roy left early in the morning, and on bad, wintry days, Winnie knew it might be after dark before he got home. She worried until he came in, tired, and said simply, "Awfully slow going today." But it was on just such alarming days that people on isolated farms found it most comforting to see Roy coming; his arrival confirmed that they still had an open connection to other people. Occasionally he would stop at the grocery store to pick up some supplies for a farmer who was running short of food and couldn't get to town. For these stranded farmers, Roy was more than a mail carrier; he was a lifeline.

Summers brought other challenges. None of his cars had air conditioning, and summer temperatures on the prairie frequently reached the high nineties or low hundreds. Rains washed out the dirt roads, and getting stuck in a mudhole or slipping into a ditch was a constant hazard. On the other hand, when dry, these same roads were worn into a corduroy surface that violently shook his car and chewed it up. Roy's 1946 Ford was followed by new ones in 1949 and 1952 and 1955, and each one endured a lifetime of rattling before he took it up to Willie Walter's Ford dealership to get a new one. In the 1950s, his car finally had a radio, and he could listen to Arthur Godfrey.

A long-running advertisement for Carl's Lunch in the *Mountrail County Promoter.*

Roy developed a routine: first he stopped at Carl's Lunch on Main Street to drink Carl Kinnoin's notoriously weak coffee and enjoy the pancakes and banter. Carl received a lot of friendly abuse for the small advertisement that he ran weekly in the *Stanley Sun* and the *Mountrail County Promoter.* For starters, it ran in the "Business—Professional" section. Given the strength of his coffee, Carl's patrons expressed doubt that he should be called a professional. They also thanked Carl for his foresight in running his ad just below those for dentist G. O. Flath, physician M. G. Flath, and a local attorney, not to mention Henry Springan's ad for tombstones—all good references to keep handy for those who ate Carl's food. But everyone agreed he made excellent pancakes.

After pancakes, Roy was off to the post office to pick up the mail. By 7:30, he was ready to run the north route. He would be back in town around 10:00, stopping for coffee and talk at Ed Brown's pool hall. Then he set off on the south route. By noon or 12:30 he would be back in town again, finished for the day. I was thrilled to ride with him, because he always found a way to make it fun. On some section roads, the grasses and weeds were clipped short between the two wheel tracks by the occasional vehicle, but along the sides of the road they grew tall, maybe three feet high, and leaned into the road, seeking more sunshine. Along sections where he knew that there were no prairie roses or other thorn bushes, Roy would dare me to stick my arm out the window and let the grasses slap my hand as we drove along. They hit you pretty hard and stung, but they never did any real damage.

He would lay his .22 Hornet rifle fitted with a scope on the back seat, and we kept an eye out for jackrabbits or a fox or even a coyote. On the day that I rode with him after getting my tonsils out, Roy spotted a jackrabbit on his side of the road, sitting in the warm sun about thirty or forty yards away. It had been nibbling at the grass poking through the melting snow, but by the time we saw it, its ears were up and paying attention. More often than not we just watched the animals, but on this day, Roy stopped the car, slowly reached back, and without making any sudden moves, picked up his rifle from the back seat. He pushed the safety on and worked the bolt to put a cartridge in the chamber, then poked the rifle out his window. He motioned for me to crawl into his lap. The rabbit was still sitting there, listening and watching. I got into position, got my finger on the trigger, and looked through the scope. I found the rabbit and placed it precisely in the crosshairs, switched the safety off, and I pulled the trigger. I heard crack, but the next thing I saw was a little puff of dirt and and the unharmed jackrabbit frantically scurrying across the field, zigzagging right and left. When Roy was playing cribbage and his crib held no points at all, he would often say, "I've got what the little boy shot at." This time, the little boy actually had something to shoot at, but the bounty proved equally disappointing. Take a lesson that I learned from Roy: squeeze, don't jerk, the trigger.

Sometimes Roy would let me drive for a short spell. There were rarely any other vehicles on the road, and we could see them miles away by their dust. When I was six, Roy put me on his lap and let me steer. When I was nine or ten, he would let me drive while he sat in the passenger seat. Once, when I was eleven, the two of us went to Polson, Montana. He drove the first two or three hundred miles, but when he got tired, he asked me to drive and crawled into the back seat and went to sleep. I drove on for several hours more, and he woke up only as I was pulling into a gas station in Browning.

I was not the only one who appreciated my dad. The *Stanley Sun* ran a regular feature called "Meet Mr. Blank: The Man of Mystery," challenging readers to guess the identity of a local person. On February 15, 1940, it offered the following clues: "Mr. Blank has a beautiful wife and four children. In his spare time he makes pillow tops and is also very handy with a hammer and saw. He has held his present job for 17 years. . . . He is on the streets of Stanley every morning at 7:30." A week later the mystery man was revealed in a story entitled, "Roy Edwards, Mr. Blank, Believes in Doing All His Tasks Well."

Starting in 1923, Roy delivered mail five days a week. His annual salary in the 1920s was $2,690. President Hoover's belt-tightening measures in 1930 reduced that sum to $2,299, where it remained for the rest of the decade, but the deflation of the 1930s meant that he enjoyed more purchasing power between 1932 and 1940 than at any other time in his thirty-three-year career. Although modest, his salary provided an adequate income, especially during such disastrous times for other families.

In 1941, however, the post office reduced rural delivery to three days a week, and this set the pattern for the next fifteen years. Roy drove his route Monday, Wednesday, and Friday mornings and was finished by noon. In a way, the change suited him perfectly. As the *Stanley Sun* had hinted, he had many other interests, hobbies, talents, and ideas, and now he had time to pursue them and turn them into additional income. The reduction in his mail-delivery schedule had cost Roy nearly a quarter of his salary, and he had a growing family to support. I was the sixth child and obviously an accident, although there were just five of us who survived because my brother Billy died before I was born. At least the timing of his reduced assignment was fortunate: before World War II, many men were looking for work, but few families had the money to hire them to build a new picket fence or put up new wallpaper. When the war came, young men enlisted and many older men left to work in defense plants, leaving few people like Roy in town. Luckily, just when he needed odd jobs to supplement his income, the community needed him to do them.

Roy was clever with his hands, could build or fix anything, and was a hard, fearless, and careful worker. He bought a small Ford tractor and a single-blade plow, and in the spring, he plowed people's gardens. He dug postholes for fences using an auger attachment. During the war and for a few years afterward, he also had a small combine that he pulled behind the tractor, so that he could help farmers harvest their wheat when manpower was so scarce. He built the first picket fence in Stanley for Archie Whitmore, and immediately five or six other families wanted one. He hung wallpaper; years later, when Mary Jensen redecorated her living room, she insisted on leaving the wallpaper up even though it really didn't go with her new color scheme, because she said that it was Roy who had hung it and he had done such a good job. He built corner cabinets and breakfast nooks and built-in bookcases. He could sew pillow tops that, according to the *Stanley Sun*, were "the envy of female needle artists." He was also the only person around who was willing to fix broken windmills. This task involved working thirty or forty feet off the ground, often in freezing, gusting winds—the kind of gusts that, if you were removing clothes from a laundry line and forgot to stand on the wind side, would whip the frozen clothes around and smack you right in the face. Roy had to get the windmill feathered and tied down before a gust hit it and knocked him to the ground, then find what was wrong and replace the broken part. He also agreed to re-tar and resurface the roof of the Dakota Drug building for Bill Eckstrom, despite having no roofing experience. As usual, he figured out how to do the job and left his customer highly satisfied. Once Bill told other building owners that Roy could put on a new roof, several asked him to do theirs, too. Roy did not agree to do quite everything— he disliked painting, for example, and avoided it whenever he could—but there was precious little that he *couldn't* do.

Roy's work was in great demand, but because his customers were also his friends, he was reluctant to charge them much. Winnie always felt that he took too little, but much of his payment came from the delight he experienced in figuring out how to do a job and from the pleased (and occasionally surprised) looks he got from his clients when he showed them his work. Once he had figured out a task and proved that he could do it, he was usually eager to move on to something new.

Roy's practical curiosity yielded its own rewards, but it never extended to intellectual topics or grand social or philosophical issues. He was never a great reader, and when it came to religion, he didn't seem to believe in prayer or churchgoing or perhaps even God. Winnie was a devout Presbyterian, and he supported everything that she and we children did in church. Every Sunday he drove us to church and gave each child a nickel to put in the offering basket, but while we were learning our Bible stories and hearing a sermon, he would head for the Ford garage, where he could chat with friends about the weather or his Saint Louis Cardinals.

Politics was another field in which Roy took little interest. Mountrail County had a turbulent and flamboyant political history; it had not only been a stronghold of the Nonpartisan League but also home to the strongest socialist and communist communities in the state. But Roy was essentially apolitical, and in our family, asking someone who they voted for was considered as nosy and impolite as asking for his bank balance. Given his long service, Roy might reasonably have aspired to be postmaster, but this position was a political appointment, and in 1946, Walt Poulsen, a disabled veteran with a high-school education but eleven years his junior, got the job instead. I used to read in the *Encyclopedia Americana* about how Franklin Roosevelt had led the war against Germany, and someone had explained to me that Democrats were in favor of government programs like Social Security and rural electrification and Republicans were not—so I was completely shocked to learn at nine or ten years of age that we were Republicans.

Year by year, Roy assumed more responsibility in the community. Although he had only a tenth-grade education, in the early 1950s he was persuaded to serve on the town's school board and then elected as its president. He was no letter-writer, but during World War II he regularly wrote to several Stanley boys serving overseas to report hometown news and cheer them up. He was also the region's official (but unpaid) local weather observer. His little white cabinet sat on spindly legs at the border of our garden, and every morning he went out early, opened the door, and took readings of the day's high and low temperatures from his two special thermometers. Then he checked the precipitation cone and carefully recorded the results. His reports appeared in the *Stanley Sun*, but those who couldn't wait for the paper would sometimes call him after a

Roy in 1944; his weather station is on stilts in the background. *Richard Edwards collection*

heavy rain or snowfall to ask, "Roy, how much precipitation did we get last night?" Each month he carefully prepared a report of his observations in triplicate for the United States Weather Bureau, where it became part of the official record.

Roy was a highly moral man when it came to his own actions and those of his family, but he was generally tolerant and non-judgmental toward others. Occasionally, however, his principles of honesty and fair play combined with a stubborn streak to overwhelm his usually sunny nature. He disapproved of some farmers who had lost their crops to drought in the 1930s and failed to pay their bills, then rebounded in the 1940s with good weather and high wartime wheat prices, but still never repaid their debts. Roy considered this behavior wrong, and that one's obligation was to repay those from whom one had borrowed. "Before you buy the new car, repay the old debt," he said. He could never quite look past the fact that in their new prosperity these farmers had dodged their old debts.

One family, the Brookses, the only Jewish family in town, owned the main clothing store. During the war their son, to all appearances a fine prospect for the military, somehow received an exemption. In the midst of writing to other town sons serving in the war, Roy thought that the exemption was wrong and refused to speak to the family. One day Sam Brooks, the father, stopped by our house, wanting to repair the breach. But Roy, meeting him out in the yard, refused to discuss the matter and brushed him off. Roy was usually free of prejudice and a stickler for fair play, so I doubt that their being Jewish was the issue. Still, it looked like prejudice; perhaps it was, and for whatever reason, Roy was unwilling to let the matter drop.

Normally, however, he was a friendly man who loved inventing new ways to have fun. When a traveling lyceum came to town, Roy and a couple of his friends put together a show featuring their tap dancing. (Roy had seen tap dancers in France, but had no training.) On another occasion, he and Ed Brown delighted an audience at Legion Hall by dressing up in women's clothes and performing a skit mimicking a women's bridge club. He invented skiing in flatland-bound Stanley by tying a forty-foot rope to the hitch of his car and towing the skier through the snow like a boat pulling a water-skier. Clarice's fiancé, Bob Meacham, a stylish young sailor from Saint Paul, visited one winter and gamely agreed to try the skiing. He was promptly deposited upside down in a snowbank in his overcoat and fedora.

Roy also liked fishing and usually made at least one trip each year with his friends, driving up the gravel roads many hours into Canada to Lake Carlyle, now called White Bear Lake. He liked to hunt and camp, as well. He was a naturally good bowler and an expert golfer on the region's prairie fairways and oiled-sand "greens." When he started, he played with only a 2 iron and a putter; later he carried five clubs in a round cloth bag. The club dues in Stanley were ten dollars a year, and there were no employees, so the nine-hole course was maintained by volunteers including Roy, who used his Ford tractor to mow the fairways. On many weekends, he traveled to Parshall or Watford City or Williston to play in the one- or two-day tournaments, and occasionally he came home with the trophy. When he won the 1933 American Legion Open in Williston, his prize was a brass pitcher.

Sadly, the golfing day that he and Winnie remembered most was a hot August day in 1932 before I was born. Their son Billy had been sick for several weeks with a series of childhood diseases and convulsions that left him weak, but early that morning, Dr. M. G. Flath had told Roy that he thought Billy had finally "turned the corner" for the better and that Roy should go ahead and play in the Williston tournament. Roy left for the tournament, but that afternoon Billy suffered a spike in his fever and another violent convulsion, black foam and vomit spilling from his lips. Winnie called M. G., but he could do nothing, and Billy died. Someone got word to Roy, who returned right away, but even so it was evening before he got home. Winnie was distraught and bitter that Roy had not been there, and Roy was devastated. Long afterwards, when she was an old woman, I asked Winnie if she ever still thought about Billy. "Only every day," she replied.

Roy also liked to travel, which for us meant going to visit relatives, such as Winnie's folks in Polson. We rarely ate in restaurants or slept in motels, instead stopping for a picnic beside the road and driving all night if necessary until we reached our destination. Roy decided to improve on this mode of travel by building a trailer that he could pull behind his car. His first trailer, built in 1931, was essentially a storage trailer for his tent, fishing gear, campstove, and food. In 1937, however, he began building a real house trailer, more like a camper than the behemoths you sometimes see on the roads today. Even with one of the new flathead V-8 engines, his Ford produced only seventy-five horsepower, so he couldn't pull anything heavy up to Polson, near Glacier National Park. So he designed a lightweight trailer with an aerodynamic shape. He had no plans to go by, except those in his head, and the result was like nothing that had been seen before in that part of the country. Townspeople kept stopping in the yard, asking, "Mind if we look in here now, Roy, and see what you're doing?" Built of wood, his trailer included two beds as well as places for storing clothes and other items and a fold-out table. The *Stanley Sun* called it "a work of art."

The family decided to test the trailer on a short run to Fargo before driving the long haul to Polson. It was quite a local sensation, and Clarice and Evelyn, then thirteen and eleven, were extremely excited. They rode in the trailer, playing cards and imagining the thrill of spending the night in it, and chatted with Roy and Winnie through a telephone line that Roy had hooked up between the car and trailer. That evening, when Roy pulled into his sister Orba's driveway, the girls were met with expressions of awe and admiration from the Asker boys, their cousins. When Roy, ever the generous guest, said, "Say, I bet you boys might like to sleep in the trailer tonight," they quickly answered, "Yes, we would!" Evelyn was dismayed. "No!" she thought. "This is *our* trailer, and Clarice and I should have the first chance to sleep in it." But she knew that her outrage was ungenerous and that no protest was possible, so she said only, "Oh, that would be fine." She was both mad at Roy and kind of proud of him.

Roy came to embody those values and virtues his Old Stanley neighbors and friends so prized: devotion to family, perpetual honesty, deep modesty, civic engagement, and a steadfastness that meant he could be counted on. He didn't begin his adult life this way, but he grew into it. He was never perfect, and he had his weaknesses and foibles. What a society celebrates, however, it tends to get, and Roy was gradually shaped by those expectations. Residents of small towns have the same range of good and bad character, of noble and nasty intent, as people in big cities—but perhaps because they lack the formal institutions that bigger places possess, small towns rely more on the steadfastness of their leading citizens. Roy became such a citizen, and his friends and his community admired the life that he constructed for himself.

Roy's last few years were less happy. In 1956, he and Winnie made a momentous and surprising decision: he retired from the postal service and they moved to a Washington, D.C., suburb to be near my sister Clarice and her husband Bob and their children. Although the move proved beneficial for me, it was calamitous for Roy. He began to lose his hearing and encountered other health problems, and his postal service pension, which had seemed so ample in Stanley, proved entirely inadequate in northern Virginia. But there was a far deeper problem. Retirement had deprived him of his role as a necessary and responsible community member. No longer a vital connector of farm families, he was shorn of his thick network of friends, activities, pursuits, and delights. He never complained, and he enjoyed the love of his family. But he had entered a different society, one in which his experience and steadfastness were not needed or valued—even, I am ashamed to admit, by his youngest son. In *The Uprooted*, historian Oscar Handlin wrote about immigrants crossing the Atlantic: "Emigration took these people out of traditional, accustomed environments and replanted them in strange ground, among strangers, where strange manners prevailed. . . . No one moves without sampling something of the immigrants' experience— mountaineers to Detroit, Okies to California." And, we might add, Roy to Virginia. He died in 1971.

DEVOTION TO COMMUNITY
· · · · · · · · · · ·
HOMETOWN DOCS
· · · · · · · · · · ·

I was eight years old when I first experienced one of the terrors of growing up in Stanley—going to the dentist. There were two dentists in town, Dr. Flath and Dr. Jensen, and naturally we went to Dr. Flath because he was Presbyterian. Dr. Jensen was a nice man but a Lutheran, so going to him would have been quite an insult to Dr. Flath. We never went for a checkup, only when a tooth hurt and needed fixing, so I knew from the start that this trip would not be a pleasant one.

Dr. G. Oakley Flath came from a remarkable family. G. O.'s older brother, M. G., was the town's leading doctor from 1911 until the 1970s, delivering nearly every baby in the county. Their uncle, Dr. Anton Flath, was a physician, farmer, sometime county coroner, and M. G.'s associate early in his career. Isabel, G. O.'s wife, was a woman of great presence who—among many other distinctions—was the Presbyterian church's long-serving choir director.

The Flaths felt a deep obligation to raise the quality of life in Stanley. Their devotion to the community stemmed not from self-promotion, but rather from an unspoken but real sense of duty to others. Each one lived this obligation in his or her own way. They served on various town boards and committees, as many residents did, but their commitment went far beyond such routine service. The Flaths—M. G. and his sister-in-law Isabel especially—organized their very lives around their aspirations for their community. Because they held high expectations for others as well as for themselves—indeed Stanley's highest expectations—they could easily have been dismissed as preachy, pretentious nosy parkers, but in fact they commanded enormous respect, a clear testament to their dedication to the good of others.

Devotion to community, the willingness to help and cooperate with others, was a virtue both expected and admired in Old Stanley. It was a duty, to be sure, but for most it was felt less as an obligation than as something undertaken gladly. It came into being when the values and traditions of the settlers met the harsh circumstances of the prairie, creating both a need for helping each other and a habit of mutual dependence and sharing. Of course, there were some who rejected this ethic and went it alone. The farmer Charlie Arndt, for example, my mother Winnie's second cousin by marriage, had a reputation for being hard to work with, tight with a dollar, and not quick to help. He insisted on driving a Buick when everyone else had Model Ts. But the Charlie Arndts were the exception; the habit of aiding others, whether in small personal matters or by supporting the community, was widespread. When settlers moved to town, they brought with them the expectation and experience of cooperation. Indeed, the feeling even influenced politics. This region spawned cooperative associations and supported the populist Nonpartisan League, and many folks even voted socialist or communist. The Flaths simply took the sharing ethic and made it a way of life.

When one of us kids had to go to the dentist, Winnie would remind us as we left the house that morning. "Now after school remember to go over to Dr. Flath's office, he's expecting you!" She never came with us, even Clarice or Evelyn when they were young, believing that if you were old enough to go to school, you could get yourself to G. O.'s office. In 1927, M. G. and G. O. had purchased a substantial brick-faced building on Main Street, later called the Flath Building. Stores occupied the first floor, and they and Doc Jensen all had their offices on the second. A visit to G. O.'s tiny office began on the

Dr. G. Oakley Flath shortly before finishing his training at the Chicago College of Dental Surgery, 1915. *Warren Flath collection*

sidewalk in front of the Flath Building. A door opened onto a long, straight, somewhat dusty stairway, lit only by the glass in the door and a single bulb at the top of the stairs, which led straight up to torment. Walking up those steps filled me with foreboding and gloom. At the top, where no space had been wasted on a sitting room, was his door with gold lettering on the half-glass: "G. O. Flath, D.D.S." I opened the door and was immediately in G. O.'s presence.

"Hi there, Ricky, how're you? Come in, come in. How's your Mom? Your Dad hit some nice drives in the tournament on Sunday. I'll get to you in a minute, just take a seat there." Oakley never employed a receptionist or an assistant, so he greeted you himself, and his welcome was invariably warm, with a big smile and perhaps a little joke. He was a talkative man—indeed, he talked nonstop throughout your visit. Perhaps he did it intentionally, to take your mind off what you might otherwise be thinking. But you knew his friendliness would not save you.

Stanley did not have fluoride in its water—the United States Public Health Service didn't start promoting fluoridation until 1951—and anyway we drank water from our own well. No one had heard of fluoride toothpaste or flossing, and our brushing was hit-or-miss. So I got a lot of cavities, in nearly every tooth behind my canines, which meant a lot of visits to G. O.'s office.

He got me properly situated in his chair, with the water swishing in the little white porcelain spitting-bowl on my left, his tray of instruments on my right, and directly ahead of me, his old-fashioned electric drill hanging on its hook. Beyond the drill I could look straight out the window. There was a visitor's chair near the door, a chair where Winnie could have sat had she chosen to accompany us, which she didn't. Because this chair was behind the dental chair, you couldn't tell whether someone had quietly entered and sat down. My sister Evelyn always felt that she shouldn't cry or make any sound while G. O. was working on her, because the next patient might have come in, and she wouldn't want to embarrass Dr. Flath.

Oakley grabbed his pick and poked around my sore tooth, talking all the while. "Well, let's see what's causing you the trouble. Oh, I see the problem now! Not too bad, not too bad! This won't be too much, not too much!" Having diagnosed the problem, he set to work, taking the drill off its hook. Its spindly arms held thin cords or belts wrapped around little wheels. The belts were connected to a small motor somewhere out of sight and delivered its feeble power to the drill bit, which was very much in sight. The whole contraption was controlled by a foot pedal. When Oakley touched it, the belts came to life, spinning slowly and making a high-pitched, pulsating whine. With the sluggishly rotating drill bit, he started grinding on my inflamed tooth. The process seemed to go on forever. G. O. used Novocain only infrequently, but he kept up his cheerful banter: "You playing baseball this year, Ricky? You a second baseman, like your brother Jack? I thought the Legion team was going to win the league last year, but they had bad luck, didn't they. They've got some nice pitching this year, though, don't they?"

Finally he hung the drill back up on its hook and said, "Well, I think that got it. Just let me have a look." Again he took up his pick and his little mirror, poking and probing the sore tooth. Inevitably he added, "Just need to get a little more now," and down the drill would come once more. When he was finally satisfied that he had drilled out all the decay, he prepared the filling. G. O. had a limp from an early farm accident, and I watched him as he hobbled over to a big cabinet with many little drawers and doors and a narrow granite countertop and began to mix the compound. "I need a little bit of this powder," he said, pouring some greyish material into his small marble mortar. Taking out a small bottle, he said, "We've gotta have a few drops of this." He continued in this vein as he added two or three more ingredients, making you feel like he was letting you in on dental secrets. He had a slight palsy, so he took the mortar in his left hand and held it tight against his chest, using his right hand to mash the concoction with the pestle. When the material was ready, he stuck it in the tooth—or, more likely, teeth—that he had excavated. When he had tamped it down and was satisfied with the result, he finally released me to fly down those long stairs, where the window in the street-level door now shone like the beacon of freedom.

I was not the only one who dreaded those visits. Clarice has similar memories, and when I asked my aunt Irene, at ninety-two, what she remembered about the Flath family, she said, "To go up to see Oakley was the longest flight of stairs I have ever seen. My knees were knocking on every step. When you got to the top, the door was right there. It was torture." Yet such was our trust and admiration for him that my parents had perfect confidence in sending us to him alone, even as small children. G. Oakley Flath would do his best for us.

His best must have been pretty good. G. O. finished his training in 1916 at the Chicago College of Dental Surgery, when modern dentistry was just emerging, and his dental equipment was primitive, not to mention scary. Prior to that time, many people, including my dad Roy, simply lost all their teeth early in life. Yet despite our many cavities, none of my siblings or I lost our teeth, even fifty, sixty, or seventy years later. It is a testament to G. O.'s early care.

Our parents were not close to the Flaths socially, but in other ways, sometimes in intimate ways, this exceptional family had a powerful effect on our lives. G. O. fixed our teeth; Isabel worked to elevate our cultural life; and M. G. delivered all five of my siblings (but not me; on his recommendation, Winnie went to the hospital in Minot for my potentially difficult birth). He tended to all of our maladies, attending Billy when he died and taking out my tonsils. He was a revered and trusted if somewhat remote figure. The Flaths were not stuck up, but they were a much more educated and high-minded family than we were, and they undeniably represented Stanley's professional and cultural elite. We were perhaps a little bit overawed by them.

M. G. was the first Flath to come to Stanley, but the first to arrive in North Dakota was his uncle Anton, who settled near Churches Ferry, some 150 miles to the east. Anton was an imposing man, taller than either M. G. or G. O., with big bones and broad shoulders. He always seemed to carry on multiple lines of work, doctoring and farming and renting out houses as though he was too big for any one occupation. In 1914, he moved to Stanley, where M. G. was already in practice, but Anton refused to set up a joint practice, so even though they treated each other's patients when one of them was away, they kept their own books and had separate offices. "He was a German," M. G. said, "and some Germans, you know, you can't convince them of anything." In addition to doctoring, Anton continued to farm his place in Churches Ferry. He had other property, as well, and he owned the big house off Main Street in Stanley where he lived with his wife and other family members. It had two full stories and a large attic and was always filled with various Flath relatives; in 1920, Anton's household numbered eight, including his eighteen-year-old daughter Marie, the youngest resident, as well as M. G. and the newly married G. O. and Isabel. Later the couple moved into their own house on Main Street, but M. G. stayed on, even buying a share in the busy house after Anton died.

In the mid-1920s, the state's attorney charged Anton with molesting a fourteen-year-old boy, George Smith. Anton vigorously denied the boy's testimony, but then the state produced evidence that two other boys, ages seventeen and eighteen, had also been molested. Anton was convicted of taking indecent liberty with a child and went to prison. In 1929, he appealed the verdict on the grounds that the boy was too old to be considered a child under the law, that the statute was unclear, and that the act complained of did not amount to "indecent liberty" because the boy had given consent. The North Dakota Supreme Court denied his appeal. Anton appealed again in 1931, however, and this time the court ruled that since the original charge had named only George Smith as a victim, the state had made a "prejudicial error" in introducing evidence about the other two boys. The court ordered a new trial for Anton, but the case was finally dismissed in 1933. Although M. G. continued to live with Anton and their names appeared together in their weekly ad in the *Promoter*, no one in Stanley ever attached any taint from the sensational case to M. G., and Anton gradually receded from public view.

Milford Garbutt ("M. G.") Flath grew up in Drayton, Ontario. In 1899, at age sixteen, he came to Churches Ferry to live with Anton, who promised to help him become a doctor if he would work for some years on Anton's farm. In 1906, M. G. began medical school at Northwestern University in Chicago. After his third year there, he returned to Churches Ferry to spend the summer. When his uncle left for a three-week vacation, he told M. G. that if any medical problems came up, "Do the best you can!" Soon M. G. was called to attend a woman about to give birth, a topic he had not yet covered at medical school. Fortunately, a midwife was there to assist him, and the child was the first of thousands of North Dakota babies whom M. G. would safely deliver. During his final year at Northwestern, he gained practical experience in delivering babies at a "lying-in" clinic in Chicago's impoverished West Side, serving women who were too poor to go to a hospital or get other care.

In 1910, M. G. returned to Churches Ferry and began looking for a place to set up his own office. He heard about a physician in Stanley who wanted to sell his practice, and in January of 1911, he bought it and became one of three doctors in town—one of whom was a horse doctor who occasionally treated people. M. G. would see patients until 1971, though he continued to go to his office until his death in 1985.

At first, the care that M. G. could offer his patients was limited. Scientific medicine was still in its infancy; doctors now understood that germs caused infection, but they had no antibiotics to fight them. Because he was competing with home remedies—kerosene on sugar for a cold, sulfur mixed with molasses for fever, a mustard poultice for inflammation, turpentine on cuts—even soap and hot water represented an improvement. The 1918 influenza epidemic tested M. G. severely, for he could provide little more than comfort to the victims. He was one of the first to contract the disease himself, falling ill in October but recovering after five days. November and December were brutal months, and perhaps a hundred people in the county died. Expectant mothers were especially vulnerable. M. G. worked night and day; he hired a driver and slept between farm calls. He came in from the country at 5:00 A.M., slept until seven or eight, then got up and started his town calls. One day he went to see a man who had acute appendicitis and fell asleep while talking to him.

Motivated by the horror of the epidemic, M. G. become a continual student. He soaked up *Current Therapy* and other journals and books that reported on medical advances, study-

Dr. M. G. Flath as a stylishly dressed young doctor in Stanley. *Warren Flath collection*

ing for an hour or two virtually every day. He attended medical conventions in Minot and Minneapolis and Chicago, and he learned from his own practice. For example, although medical school had taught him to move pneumonia victims near a window because they needed fresh air, experience quickly taught him that patients in closed rooms did much better. Even so, until the advent of antibiotics, pneumonia was a killer that M. G. could do little about, and tuberculosis was a virtual death sentence. His primary weapons were soap and hot water, boiling his instruments, and offering comfort and even spiritual support.

M. G. made a striking appearance, although he was not overly handsome by traditional standards. Of medium height, slight and trim, with erect posture and beautiful silvery hair, he always dressed in a well-tailored, almost elegant suit. He exuded an air of unpretentious sophistication not usually associated with a country doctor. People often gave him a second look—especially women. Yet he never married, which puzzled Irene: "I can't imagine how he escaped. . . . He was so good-looking, such a nice man!" He did have a sweetheart, although she was conveniently unwilling and far away—a girl he had gone to high school with in Churches Ferry, who had refused to marry him, according to the story put about, because she needed to care for her mentally defective brother. Eventually she became a teacher and moved to Saint Paul, where M. G. would visit her occasionally, but apparently he was not an ardent suitor. Today such behavior would raise suspicions that he was gay, but most people in Stanley simply thought it lucky. If he had married, perhaps he would not have been so dedicated to his doctoring.

M. G. did have the normal Flath affliction, a highly positive attitude toward life. Although at first his medicine was as primitive as G. O.'s dentistry, his nature was to be cheerful, competent, and reassuring, and his willingness to go anywhere in the county to treat his patients brought great comfort to rural residents, especially Mountrail's expectant mothers. M. G. charged nineteen dollars for a delivery, which included two visits: the delivery itself and a follow-up visit two days later. In the 1910s and 1920s, he cleared about $2,500 or $3,000 a year, a modest amount for a doctor—indeed, only a few hundred dollars more than Roy earned for delivering the mail. Perhaps he made more on paper, but most people expected him, like a merchant, to carry their bills on credit until the fall, when the crops came in. Even then, many did not or could not pay. Nevertheless, he had a well-known policy of treating everyone regardless of whether they could pay or not. One farm family did not pay him a nickel for seven deliveries, and still he went out to perform an eighth. "I never refuse to go!" he declared. "Some doctors would refuse to go, but I never refuse anybody. When they call me, I go day or night. I wouldn't say, 'Wait till morning,' or something. I go right away!"

One man who had never paid M. G. for his services during the 1920s came back to his office in 1974 and asked, "How much do I owe you?"

"Well, I don't know. It's so long ago, I haven't kept track of it."

"Well, I know. I owe you three hundred and fifty dollars."

The man promptly pulled out his checkbook and wrote M. G. a check, roughly fifty years late. He was ninety-one, and perhaps he was wondering how he would explain his delinquency to his Maker.

With the drought and the Great Depression, M. G.'s income plummeted. There were fewer babies to deliver, fewer people called the doctor, and fewer paid him even if they did call. He lost money when the banks failed and had to borrow two thousand dollars to stay in Stanley. He reduced his fee to nine dollars per delivery, but even then he could hardly keep his practice going. "I was on twenty-four-hour call all the time," he said. "I was so hard up I just had to work all the time I could work. When I wasn't working I studied." His fortunes improved only after World War II began.

I had little direct experience with M. G., because our parents called him only in dire circumstances. No one thought about annual checkups, and we weren't required to produce medical forms for school or to play sports. For ordinary diseases, like chickenpox, mumps, and measles, there was no need to call M. G. because Winnie knew just what to do. She'd close the drapes to keep the room dark, tuck us in bed, check our temperature, ply us with liquids, and put a quarantine sign on the door: "Measles—Don't Come In." M. G. was called several times when my brother Billy was sick with whooping cough, but that was before my time.

Roy saw a lot more of M. G., because in the winter he would sometimes call Roy to drive him out to some remote farm to deliver a baby or to patch up an injury, or because someone was near death or had just died and needed a death certificate. When he chose not to use Roy, M. G. was an intrepid driver in his own vehicle, crossing miles of prairie by himself to find some isolated farmhouse. Such calls often came at night, and he would go regardless of the time or the weather or the condition of the roads.

In the early years of his country doctoring, a team and sled was the most reliable means of getting through the prairie snowdrifts, but despite growing up on an Ontario farm, M. G. truly disliked horses and wanted as little to do with them as possible. Instead of owning his own team, he usually rented a team at Square Warren's livery stable. These teams were sometimes difficult to control, and more than once they ran away with him. To bring them back under control, he would either run them off the road into the snow, which would slow them down, or run them up against a barn, which stopped them. On one occasion, when his team got back within sight of Square Warren's stable, they bolted and began to gallop towards home. One horse went through the stable doorway, the other continued outside, and they demolished the sled. M. G. had jumped to safety at the last minute.

Understandably, M. G. really liked the idea of a "horseless carriage," and he began using an automobile in 1911. His first vehicle was a two-cylinder car with no windshield, no top, and no transmission, powered by a chain drive on each wheel and featuring acetylene-gas headlamps. Three years later he bought a little Buick roadster with electric lights and twenty-two horsepower, but it was a ramshackle contraption. One day as he was driving, a front wheel came off and rolled across the prairie. By the 1930s, he had switched to a more reliable Viking. One Sunday afternoon he was called out to see Bert Aas on his remote farm way out southeast of Palermo, and he asked his friend Lyle Reimers to ride along. The road goes over some rolling hills, and M. G. decided to speed up to give Lyle a few butterflies in his stomach as they crested the hilltops. One hill was steeper than expected; M. G. lost control, and the Viking jumped the ditch and wound up out on the prairie, leaving both men shaken and disheveled. They got the car back on the road and made the rest of the way to the Aas farm without incident, but Mrs. Aas was shocked to see them, especially the usually immaculately dressed M. G. "You look like you been in a fight!" she scolded.

M. G. put twenty-five thousand miles a year on his cars just driving across the countryside. His early vehicles either had no heater or only a feeble one, so in the winter he had to bundle up against the cold in a big sheepskin coat and cap, covered over with a heavy blanket that had holes for his feet. It was in the 1920s that he began asking Roy to take him out in his improvised snowmobile whenever he faced especially bad conditions, such as high snow drifts or heavy mud. M. G. rarely knew whether he would be paid for such visits, and Roy certainly received little or nothing beyond the cost of the gas, but they both went because they were needed.

M. G. developed some routines to refresh himself. Nearly every day around noon, he went into Bill Eckstrom's Dakota Drug on Main Street, said hello to Bill and to Arne Springan when he worked there, and strode directly into the back office. He sat in Bill's chair and napped for fifteen minutes, emerging refreshed and ready for the rest of the day. On Sundays he played golf. M. G. had been one of the founders of the Stanley golf course in 1925, and with G. O., Henry Springan, and Harold Borg, he formed a foursome that played together for decades. G. O. played despite his limp, Henry was missing three fingers on his left hand, and M. G.'s handicap was a ferocious slice on his drive. As a practical man, however, M. G. decided to adapt to the slice rather than struggle to correct it. Stanley's first hole ran alongside the Great Northern tracks, so on the tee, M. G. aimed directly for the railroad. The ball would soar out towards the tracks, looking as if it would surely land out of bounds, then reliably slice back in and nestle into the sparse, dry grass 150 yards out in the middle of the fairway.

Like G. O.'s dental office, M. G.'s medical office had no assistant or receptionist. M. G. relied on the relatives he lived with in Anton's house to be his answering service. Few of his patients remember being in his office, for he did almost all of his doctoring in the homes of the pregnant, injured, or sick, especially before the community hospital was built in 1952. Along with tonsillectomies and appendectomies, delivering babies was the biggest part of his practice, and he delivered nearly thirty-three hundred babies before he retired. He even kept a portable table in his car, so if a woman's own bed was dilapidated or unsuitable, he could unfold his table and she could give birth on it.

M. G.'s relationship with midwives was complicated. He depended on a few of the better-trained ones as auxiliaries, but others were ignorant and more trouble than help. One midwife whom he respected, Mrs. Hoffman, had been trained in Norway and worked south of the nearby town of White Earth. One time M. G. was called to a ranch for an imminent delivery, and when he arrived, Mrs. Hoffman was already there preparing the mother. Examining the woman, he said, "I think she's going to have twins." "Bet you a nickel you're wrong!" Mrs. Hoffman rejoined. After the first birth, though, sure enough, a second baby presented. Of course, it was hardly a fair bet: M. G. had a stethoscope.

Other midwives were less helpful, and often M. G. was called in after a midwife ran into trouble. The first Mountrail County baby that M. G. delivered was born on a farm south of Ross. A midwife who had been drinking whiskey in the kitchen with the family greeted him when he arrived.

"This woman cannot have her baby," she said. "You'll have to take it."

"How do you know?"

"Well, I examined her."

"*You* examined her?"

M. G. saw that her hands were extremely dirty. He told the husband, "I don't want to be responsible for any infection here. I want you to understand I'll be careful as far as I'm concerned, but if this woman has previously infected your wife, I don't want to stand the responsibility." After eight or ten hours of labor, the woman delivered a normal baby. M. G. went back two days later to check on her, and she was doing fine.

On another occasion M. G. attended a confinement south of Stanley, where an old woman was serving as midwife. "This woman can't have her baby," she said. "You'll have to take it. I've done everything I can for her, but she can't have it."

"Well, what have you done?"

"I greased her from her waist to her knees, and it hasn't helped a bit."

"Well, the first thing we're gonna do is get a great big dish of hot water and soap and give her a bath. Let's get all this grease off her."

One complication in M. G.'s practice was the great ethnic and religious diversity of Mountrail's settlers—Norwegians, Syrians, Finns, Mennonites, Germans, Bohemians, and others—many of whom spoke little or no English. Because he traveled alone, he often had to ask the husband or another family member to assist in the delivery, especially during extractions. While attending a Finnish woman in a house where no one spoke English, he enlisted her husband to administer the chloroform, using hand signals or a shake of the head to indicate more or less. "You saved my life!" she told him through an English-speaking relative who arrived after a successful extraction. "I had so much pain I thought I was gonna die."

M. G. also became county coroner and had to deal with Mountrail's dead, crisscrossing the county to examine corpses and issue death certificates. One day he got a report that a body had been discovered in a ditch, the man having been dead for three days or more. To help him, he enlisted Henry Springan, who ran the town's mortuary in the back of his furniture store and did a brisk side business selling tombstones. Together they had a fine spring outing on the greening-up prairie as they brought the body back to town. M. G. often used his post-mortem examinations to determine whether his diagnoses had been correct. One Finn from Belden came to Stanley for treatment, lodging in Mrs. Munson's nursing home. M. G. examined the man and told him that he needed surgery for an intestinal obstruction. "No, I don't want surgery," the man said, and he refused to permit it. Returning a couple of days later, M. G. found that the man had gotten worse. "I want to be taken to Minot to have the operation," he said. "There's no reason to take you to Minot," M. G. responded after examining him again. "You're gonna die." And the next morning, he did. The doctor did a post-mortem with Henry Springan over at the mortuary. They found that the man had peritonitis and holes in his intestine but could not find his stomach anywhere in his abdomen. "Well, he must have a stomach," M. G. told Henry. "Let's open up his chest." They split open the ribs, and M. G. removed the trachea, heart, and lungs. Behind the left lung was the man's stomach. M. G. had wondered why he had such a thick chest, and there was the answer. The man was eighty-nine when he died.

Medical care in Mountrail County changed when the new community hospital opened in 1952. There had been Mrs. Munson's nursing home and Mrs. Johnson's maternity home—really just private residences—but for all other treatment Dr. Flath came to you, and before he left, you pulled out your wallet and paid him. The custom of house calls did not disappear immediately after the hospital opened, because no one had medical insurance and people were unused to paying for a hospital room when the doctor would come right to your house. Gradually, however, the practice changed, until the idea of a doctor coming to your house seemed as improbable as paying the doctor yourself in cash.

M. G.'s commitment to being Stanley's doctor came from choice rather than lack of opportunity. As a graduate of Northwestern, he was vastly better educated than the usual run of North Dakota doctors. In 1910 he visited the Mayo Clinic, then just a small building with three doctors but already known for the advanced techniques being developed there, and he watched Dr. William Mayo perform pioneering gallbladder surgery. During his career he served for a short time as an anesthetist, performed induction examinations for the army, consulted for the Great Northern Railway on injury and accident cases, and kept up with all the recent medical advances. Relocating to Minot or Fargo or even Minneapolis would have meant access to a hospital, a more regular life, a fancier clientele, a much higher income, and no perilous night trips to remote farms. But M. G. saw how much Stanley needed him. "I never got tired enough [of doctoring in Stanley] that I wanted to quit," he said. "I just got more interested in my patients. I felt the duty of doing the work. After I got started [in] practice here, I wanted to be loyal to the people here."

M. G.'s enduring devotion to his adopted community made him a local hero, but in a sense he was an ordinary part of the town, too. All the leading citizens were expected to contribute their time and energies to the community, often at personal cost and rarely to gain personal accolades. Not everyone lived up to this standard, Banker Nelson being one who wasn't always greatly appreciated, but M. G. exceeded it. His unceasing service regularly put his own life at risk and earned him only a modest income, and each selfless act was typically unknown except to the family being served. Only later, when nearly every family had its own heroic M. G. story, did it begin to dawn on people that his ministrations were something special.

That day when I was eight and sitting in Oakley's dental chair, I waited for him to finish my new filling and release me. Finally he said, "Okay, Ricky, you're all done! Say hello to your Mom for me!" and I went tripping back down those long stairs. I was the last of a long line of Edwards children whose flaws G. O. and the other Flaths worked to fix. Their service to our family was but a small part of their profound and lifelong commitment to help everyone in town. Regardless of opportunities elsewhere, the Flaths chose Stanley and earned our gratitude for their lives of relentless devotion to community.

PLUCK

· · · · · · · · · · ·

SISTERS AT WAR

· · · · · · · · · · ·

With the arrival of World War II, the Great Depression finally released its grip on North Dakota, but even greater opportunities arose elsewhere—unheard-of opportunities. By 1943, word had spread to Stanley that the war plants out west had lots of jobs, even for women, and that they paid big money. My grandparents, Bill and Frankie Burlingame, and their son Floyd and his wife Alice had already moved to Portland, and my sisters Clarice and Evelyn heard about the possibilities from them. Bill, Floyd, and Alice all got jobs in the Kaiser shipyard. Alice worked as a nurse, and Bill and Floyd worked the swing shift out in the yard and loved it. Frankie stayed home, wallpapering and scrubbing floors and washing and canning and cooking and sewing, convinced that it was quite improper for a married woman to *work*.

Clarice and Evelyn wanted to go to college, but they had no money, and Stanley jobs didn't pay much. So in the summer of 1943, eighteen years old and just graduated, Clarice set off for the shipyards with that *pluck*—a mix of boldness, desire for adventure, and determination—that was in the Stanley air. She would work, save her wages, and wait for Evelyn to graduate the following year so they could go to college together. Clarice wanted some adventure, and going to Portland would be her first real opportunity to get out of Stanley. Her friend Betty Anderson was in the same frame of mind. Betty had graduated in 1942 and started nursing school in Minot but had run out of money; on her final day there, she used the last of her toothpaste and realized that if she bought more, she wouldn't be able to pay the bus fare back to Stanley. The two plucky young women decided to embark on their adventure together. A year later, Evelyn joined them, climbing aboard the westbound Great Northern train on the evening of her graduation. My sisters are nineteen and seventeen years my senior, and although they couldn't know it at the time, their nervy adventure would have a tremendous and positive impact on my brothers and me.

Pluck was a much-admired quality among Old Stanley people. It combined gumption or moxie with a lighthearted and daring spirit that was ready to embrace new circumstances and possibilities. The homesteaders' original migration to this cold, barren prairie had required pluck, fearlessness that overlooked the dangers and hardships and instead saw adventure and opportunity. In O. E. Rølvaag's great novel *Giants in the Earth*, the lead homesteader, Per Hansa, had pluck to excess, but his wife Beret's complete lack of the trait doomed their enterprise to failure and tragedy. Town residents, too, admired pluck, the kind of spunk and backbone that led young Roy to develop his dodging-monkey carnival booth and his brother Swede to set off for Chicago to seek his fortune.

Roy and Winnie raised no objections to Clarice's departure, expressing little anxiety about the dangers of industrial work, the rough language of shipbuilders, or the possible temptations facing an adventurous young girl in a port city filled with sailors. They knew that she had relatives there to keep an eye on her, but there was more to it than that. When first Clarice and then Evelyn graduated from high school, Roy and Winnie felt that their responsibility for the girls was over. They were now grown up. They were not washing their hands of the girls; they simply believed they were now adults who should make their own decisions. Whether they went to the shipyards or to college, the girls knew that they were launched into the world. There would be no coming back home, complaining, "you didn't send any money." That just wasn't a thought in anyone's head.

Clarice and Betty's first stop was Spokane, Washington, where they joined a third friend, Arlene Hungate, and stayed with her parents. Arlene's father had once been the Mountrail County sheriff. The girls worked briefly in a local Boeing parts plant. They attended a war-bonds rally featuring Gene Kelly, and the girls felt good that they were contributing to the war effort, too. Soon they had a chance to see Winnie, for Frankie had suffered a serious stroke, and Winnie, heavily pregnant with me, was going out to Portland to see her. When her train stopped in Spokane, Clarice met her on the platform with a new friend in tow—a sailor. Perhaps Clarice was showing off her new freedom.

Clarice and Betty soon moved on. Portland, Oregon, and Vancouver, Washington, lie on opposite banks of the Columbia River. Before the war they were small cities serving as commercial centers for the surrounding lumber, fishing, and ranching communities, but the war transformed them into industrial hives. Between January 1941 and March 1942, the Kaiser Shipbuilding Company built two large shipyards in Portland and one in Vancouver. Together with three existing yards, they employed nearly a hundred thousand workers. Kaiser quickly exhausted the local pool of male workers, and even high wages and wide recruitment efforts could not bring enough men to the shipyards. Needing more labor to fulfill the Roosevelt administration's wartime production goals, Kaiser turned to women. It was a controversial decision, but the company had little choice.

Soon after they arrived, Clarice and Betty went over to Kaiser to apply for work. The girls realized that the company must be desperate for workers, because they were quickly hired—as *welders*. Neither one had done any welding before—indeed, they had never even heard of it. But the man told them that it was the only open position. Eager for work, they replied, "We'll do it!" Immediately they were enrolled in a training program, and the girls were delighted. They couldn't believe that they were going to be paid to *learn* to weld! When they returned home that evening and told Bill, Frankie, Floyd, and Alice about their new jobs, all four of them were unhappy. "Welding is way too rough a job for girls to do," groused Frankie. Clarice and Betty refused to be dissuaded; they knew that they could do it. And though they had no money to buy the hoods, hard-toed boots, and special leather suits that the job required, the shop selling welders' gear let them pay on credit out of the plump paychecks that waited in their future, and they began their training. They would spend much of their time in these leathers and boots, wearing them from home to work, at work, and back again five days a week.

In peacetime, ships were built by skilled craftsmen—ship-fitters, sheet-metal workers, machinists, and electricians who had undergone years of apprenticeship and belonged to exclusive trade unions open only to white men. Faced with the new problem of building ships with mostly unskilled workers, Kaiser transformed this craft-based process into something akin to mass production. Because ships are too big to move down an assembly line, they had to remain fixed on the ways while teams of workers moved from task to task around and on and in them. Kaiser's first key innovation, sometimes called "parallel production," was to prefabricate many of the ships' parts in shops scattered around the yards. These preassemblies were then brought out to the ways to be put together. First the keels were laid down, then came a steady stream of other components, including bottoms, bulkheads, engines, deckhouses, and steel plates for the decks. Each part was lifted onto the embryonic ship by huge cranes, or "whirlies," and fixed in place. Limited space on the ways had always been the bottleneck in ship production, but Kaiser's system changed that. It reduced the amount of time a ship needed on the ways, because so much work could be done beforehand in the fabricating shops. The workers on the ways had only to fasten the preassembled pieces together to make a ship.

"Whirlies" lift a deckhouse preassembly to be welded onto a
Liberty Ship hull at the Vancouver shipyard, 1944. *United States
Maritime Administration*

Clarice and Betty owed their new jobs to Kaiser's second big innovation, the use of welding instead of riveting to fasten the ship's pieces together. Some consider welding to be the primary reason why ships were built so much faster in World War II than in the previous war. A contemporary issue of *Fortune* noted, "In the first world war welding was a repair tool. . . . In this war welding has made one of the greatest contributions to the speed and size of U.S. production." Riveting required time to align the steel plates, drill the holes, then set and drive home the rivets. Workers then had to grind the exterior rivet heads flush with the hull so they would not slow the ship at sea. By contrast, welding the plates together produced equally strong joins but was vastly simpler and quicker. "One welder can do the work of a four-man riveting team," *Fortune* claimed, making possible the speedy production of thousands of ships of all types. Despite rumors that welded ships cracked under stress, many welded ships survived torpedo hits that would have been fatal to riveted ships. Using parallel production and welding, whole ships could now be assembled from prebuilt subassemblies and welded together in a matter of two or three weeks rather than months.

The new methods created a huge demand for welders just as millions of men marched off to war. Before the war, welding had been a skilled trade requiring a multiyear apprenticeship, a system designed both to organize training and limit the number of welders competing for jobs. Now, however, Kaiser needed a huge number of welders and couldn't wait for trainees to emerge from the apprenticeship system. Because Kaiser's welders needed to know only a few specific types of weld, they could be adequately trained in two or three weeks. As soon as a trainee mastered the required welds, she was tested, certified, and sent into the yard.

During training, each young woman was placed in a dark little cubicle to practice her welds; the instructor would come by from time to time to show her a new weld, see how she was doing, or criticize her work. Augusta H. Clawson, another woman who worked in the yard, described what she and the other trainees were learning: "At times the whole process seems like something out of a fairy tale. Take two pieces of scrap iron, place them close together, guide the rod of metal with fire glowing from its tip, and draw it gently along the edges of these two plates and, *presto*, suddenly they are one! The terrific heat melts the metal rod and the edges of the iron plates and all the molten metal flows smoothly together. The Fairy Godmother and her magic wand have nothing on us."

Clarice and Betty learned three kinds of welding: vertical, horizontal, and overhead. They must have impressed their teachers, because overhead welding was considered the most difficult of the three, and it was not part of the standard training course. Until they had mastered all three types, however, welders were restricted to the fabricating shops and could not go out on the ways to work on the ships themselves. In a couple of weeks, the two girls obtained their certifications and became official "triple-plate Navy welder[s]," joining a growing force of women welders at Kaiser's Vancouver shipyard, the area's biggest. By October 1943, Kaiser employed over five thousand female welders, making up 35 percent of the welders in its three yards.

The Vancouver yard specialized in escort aircraft carriers, and for the next year Clarice and Betty would work on these "baby flattops." Designed to protect convoys and fight submarines, they were smaller than the behemoth fleet-type carriers but huge nonetheless, fully 490 feet long and with high decks. In 1943, Admiral Ernest J. King, the chief of naval operations, challenged the Portland-area shipyards to speed up the production of escort carriers, declaring that it was vitally important to an early victory. Edgar J. Kaiser, who ran the Vancouver yard, responded by proclaiming, "18 or More by 44!"—a colossal goal.

Workers on the ways were organized into work crews, each directed by a foreman. A crew included several shipfitters, welders, and other workers, maybe ten or a dozen in all. A crew

An escort aircraft carrier is launched into the Columbia River at Kaiser's Vancouver Shipyard. *United States Maritime Administration*

would be assigned a certain task, like attaching deck plates or welding down a bulkhead preassembly. Upon completing the task, the crew would go on to a different job. A welder might work under the direction of the shipfitters or she might be assigned a job to do by herself, but always with the same crew.

Arc welding produces a join by creating a powerful electric flash between the electrode and the material being welded, bonding the momentarily liquid metal and usually producing a shower of sparks. Each Kaiser welder needed to be clad all over in leather, from hard-toed boots to a special hood, so that every part of her body was protected from flying sparks and molten metal. She wore a thick belt around her waist hung with various welding tools. The most difficult component to handle was the "lead," the thirty-five- or forty-foot-long electric cable connected to the machine that generated the current. Everyone complained about how heavy it was. As *Fortune* reported, "The welder . . . must carry heavy rod boxes and thirty-five pound leads . . . up and down ladders, in and out of narrow passages; and he [or she] must keep on welding in rain or snow or blazing sun." At the beginning of her shift, a welder pulled her lead to the job site, got familiar with the rods she would use there, and got her heat or spark set correctly. Then she was ready to weld the rest of the day.

It was a strange and intimidating new environment. Another welder, Lue Rayne Culbertson, said she cried herself to sleep her first week in the shipyards, thinking, "What have I done, how did I ever get into this mess? I mean sure, you're going to be a welder, you get a little plate and you weld it, and then all of a sudden you go out there and see a big ship with all these cranes and operations and whatnot, and noise—it's just hard to imagine." The work was almost unbearably noisy because of the "chippers," workers who tested each weld and used power-driven steel chisels to chip out any faulty work. The noise was deafening, like many jackhammers going all the time. Augusta Clawson wrote of how "there is nothing in the training to prepare you for the excruciating noise. . . . There are times when those chippers get going and two shipfitters on opposite sides of a metal wall swing tremendous metal sledge-hammers simultaneously and you wonder if your ears can stand it. Sometimes the din will seem to swell and engulf you like a treacherous wave in surf-bathing and you feel as if you were going under. . . . It makes you want to scream wildly."

Still, the noise was not the most taxing part of the work. Welding on the ways was physically grueling and so hot that the workers took salt pills to keep from fainting. The noise level, the heat, the climbing up and down, the work at great heights—all these factors combined to make the job strenuous and stressful. A shift lasted a full eight hours, and always in the background was the notion that any worker who was lax or slow or careless was undermining "the boys" who needed the ships to fight the enemy. Sometimes a welder lay on the floor and worked while another welder spattered sparks down on her from the ceiling, and the chippers, "like giant woodpeckers[,] shattered our eardrums." Clarice's fear of heights added to the discomfort she felt working on the ways. But while welding was arduous and stressful, she was eighteen, plucky, and used to being a hard worker, and she got on with her job.

It was dangerous work. The hazards of shipbuilding included cranes lifting massive sections of steel and cable into place; decks and passageways cluttered with hoses, pipes, leads, and dangling cables; and workers climbing around on heights and in narrow spaces, using tools that were heavy, hot, and unforgiving. The environment was busy, deafening, and filled with fumes and sparks and metal splinters. A 1944 survey found that shipbuilding had the highest rate of disabling injury of any war-production industry. Another study found that in nearly every department, women experienced higher rates of disabling injury than men.

Welding and burning operations had the highest accident rates of all. The heat created by the arc at the end of a welder's electrode could climb as high as six thousand degrees Fahrenheit, nearly as hot as the surface of the sun. Flying sparks and hot slag could quickly burn through clothing. On hot days, welders found their hoods to be instruments of torture, but the risks of relaxed diligence were unthinkable. "We had to be careful we had our gloves and leathers on at all times," Betty remembered. One day she burned herself on the wrist. "Oh, that was my own fault, I guess. Apparently I got the rod too hot and didn't have my glove on tight enough." She kept on working, putting a bandage on the burn after she got home. The worst hazard was to the welders' eyes. "It was easy to have our hoods up and go to start welding before putting your hood down," Betty said. But welding for even a short time without the protection of the dark lens mounted in the hood could burn a welder's retinas and permanently damage her eyes.

Other hazards of shipbuilding turned out to be almost comically overstated. Soon after starting work, Clarice and Betty were told that they had to join the union or they would lose their jobs, and that not attending union meetings was cause for discipline. The girls didn't even know what a union was, but they did as directed and signed up. On the day of the next meeting, they went over to the union hall in Vancouver, which turned out to be a big building plastered all over with union signs. The doors were standing wide open, and rows and rows of chairs were lined up inside, all empty. "Gee, we must be awful early," Betty said. They sat down in the back and waited. And waited. After an hour, they finally got up and left, having seen not one other person. Their first and only union meeting had drawn exactly two attendees.

Many foremen were skeptical of women welders, and the girls worried about whether their welds were good and strong. Clarice had a friend who had once made a poor weld, and the chipper came over and sneered, "No wonder the ships are falling apart out there!" implying that inept female welders were causing the death of sailors. "Oh, I felt so terrible!" she told Clarice. "I had a vision of all these servicemen tumbling into the sea because of me." She made sure it never happened again. Clarice herself never suffered the humiliation of a chipper digging out one of her welds, but she understood her friend's concern. Once, as she was fastening down the furniture in a ship's hospital room, she thought, "Oh my god, what if this doesn't hold?" As the ship pitched and rolled in heavy seas, "if my welds don't hold the chairs down, there might be wounded or sick men smashing against the walls." Such moments of doubt were rare, though; she had confidence in her work, partly because she worked under a supportive foreman. Betty's foreman, too, was helpful. About sixty years old, he reminded Betty of her grandpa because he spoke with the same heavy Norwegian accent. He checked her work carefully when she started and continued to look out for her as she gained experience.

Three women welders attach deck girders at the Vancouver
shipyard, 1944. Their "leads" or electric cables are on the deck
behind them. The Oregonian

Despite their fast-growing skill, however, the employment
of women in industry was generally seen as only a temporary
and unfortunate necessity of war. As one shipyard newspaper, *Flat Top Flash*, put it, "Women in ordinary times do not
belong in shipyards any more than men belong in foxholes."
Women in the yard endured a certain number of catcalls and
leering comments despite the hugely unflattering leather suits
they wore. As historian Amy Kesselman has pointed out, the
stories and cartoons in shipyard newspapers in 1941 and 1942
stressed two features of women: their supposed ineptitude as
workers and their sexual attractiveness. By early 1943, when

the number of women working in the yards rose into the thousands, the image of the inept female largely disappeared and the image of the frivolous sexpot moved to the foreground. "With protruding hips and breasts, she appeared repeatedly in cartoons . . . reminding men and women alike that underneath the welding helmet there were creatures irrevocably different from men—provocative, jealous, nagging wives and girlfriends, obsessively concerned with their looks, addicted to gossip and eternally sexual."

Rumors circulated that young women on other crews welding deep in the hull had run into some "trouble" with male welders. One day Betty was assigned a job far below deck, and she descended into the dark, pulling her heavy lead and feeling claustrophobic as usual, finding her way by the beam of a small electric flashlight attached to her hood. Then, suddenly, her light blinked out, leaving her in the pitch dark. She struggled to find her way back, and fortunately a coworker found her and told her foreman what had happened. The assignment had badly frightened her, but it was out of the question to refuse a task. Both she and Clarice resolutely worked on their assignments, whatever they were and wherever they took them.

In 1943, expecting an increase in the number of women employed in shipyards, the Women's Bureau of the United States Department of Labor sent investigators to forty-one shipyards to study women's working conditions and make recommendations for improving their lot. Their first recommendation didn't concern women's difficulties in handling the heat, fumes, or physical rigors of the job; instead, it exhorted yard management to "secure the cooperation of men supervisors and workers." Most of the men in the yard were in their thirties and forties at least, since younger men had gone into the military, and the report noted that a resentful and uncooperative male workforce could sabotage the women's work. The report echoed what Clarice's friend Helen experienced. "None of the men would ever help us with any of the hard work," she said. "I think their attitude was that we were getting a man's pay, so we would have to do a man's work." By contrast, Clarice found that "in general people were kind to the women in the yard and wanted them to do well."

Women were not the only workers who sometimes struggled against prejudice. The yards were so desperate for workers that Kaiser relaxed restrictions on hiring African Americans, as well. Blacks usually worked in the dirtiest and lowest-paying jobs, as sweepers, helpers, and tank cleaners, but a few made it into skilled trades. Except for the communist-influenced United Electrical Workers, unions for skilled workers were segregated, and Kaiser fired black employees who refused to join the "correct" union. When blacks appealed workplace mistreatment to the management or to the federal Fair Employment Practice Committee, they were often simply ignored—or worse. Some white workers, coming from small towns and having little experience with blacks, felt uncomfortable interacting with them. One welder Clarice knew didn't want to ride the packed buses. The black people "didn't like the white gals on the bus," she said. "They would elbow us. . . . I was scared to death." But Clarice, working alongside African Americans for the first time in her life, embraced the new kinds of people and experiences she was having, laying the groundwork for her lifelong passion for equal rights. She felt there was a good spirit on the ways, which helped make the hard working conditions tolerable. "I really liked the big wages and the feeling that I was doing something good for the war, the sense that I and the others were contributing something. This job was a long way from serving malts and sundaes in Gray's Creamery [in Stanley]."

The welding itself was satisfying work. "Down on our hands and knees, or sitting on the deck," Betty observed, "when we finished our weld, we would look at it and feel good, like we had really accomplished something." Fellow welder Billie Strmiska agreed: "It was such a pretty sight under those hoods, really, welding is a beautiful art. You could sew the prettiest seams, especially those verticals, you could just rock those in there and make the prettiest seams. And when you chip that slag off and see that gorgeous weld underneath, no undercuts, no nothing, just a beautiful weld—it was a real, real satisfying job."

Their hard work paid off. Vancouver built a total of 125 ships during the war, including fifty baby flattops. The company continually urged greater production, faster work, and less turnover, and admonished against the evils of absenteeism and slacking off at the end of a shift. In 1942 it had taken over two and a half million man-hours to construct an escort carrier; by 1944 that figure was down to about one and a half million. As productivity improved, the Vancouver yard was able to exceed its goal of "18 or More by 44" by launching nineteen baby flattops before the end of 1943.

Surprisingly, the welders only rarely got to see a ship launched. One day someone pointed out the preparations for a launch to Betty, but she didn't know what ship it was or whether she had worked on it. Shipyard workers were generally kept in the dark about their work; Clarice never even knew the names of the ships she worked on. "We weren't told, and we never thought to ask," she said. The only time she learned the name of one of "her" ships was when someone gave her a picture of the U.S.S. *Midway*, a baby flattop that had come in for repairs, and told her that she had worked on it. She thought, "That's wonderful, that we were able to do that." The *Midway*, renamed as *St. Lo*, returned to service and was sunk by a kamikaze plane off the Philippines on October 25, 1944.

When they first arrived in Portland, Clarice and Betty moved in with Frankie and Bill in their tiny rented house in Vanport, Oregon. Vanport was a huge company town newly built by Kaiser, which had recognized that Portland and Vancouver were too small to house all the workers it needed in the yards. At its peak, Vanport accommodated forty-two thousand people. It was hastily constructed, sprawling, and often inconvenient. Most apartments had a living room with a kitchenette, a bathroom with a shower, and one bedroom; in many units cooking was done on two-burner electric ranges. Fifty-six units shared a laundry room and bathtub. The streets were unpaved, creating a muddy mess in the rainy season. The entire city would later be swept away in a 1948 flood.

Although space was tight, Frankie and Bill welcomed the girls, and Frankie packed lunches for them, including special treats like her homemade apple pie. Soon the girls got their own little apartment in the nearby Ogden Meadows section of Vanport, where they cooked on a stovetop and washed their clothes in a big sink, hanging them to dry on a string outside. They made their own lunches, packing their lunchboxes with a hard-boiled egg, a banana or an apple, and a container filled with milk. The tiny apartments all looked alike. One day Clarice walked into her apartment with the familiar bowl of bananas on the table but was surprised to find no one home. On closer inspection, she discovered that she had walked into the wrong unit. The sameness of the apartments was a great leveler, which Clarice liked. No matter what circumstances a girl came from, her place was just like all the others.

At first the girls rode the bus to work, but it was crowded and disagreeable. Soon four of them found a ride together with Slim, a friendly, older Oklahoma man who owned a little roadster. He liked the girls and made each one a pretty silver-colored bracelet out of a welding rod. He stopped by their apartment every morning to pick them up and brought them home in the evening, a great convenience for which they paid fifty cents a week. Two girls sat outside in the rumble seat, and the other two inside with Slim. The seat next to Slim was the worst position, because he had a way of accidentally letting his hand touch their legs, so they carefully rotated who sat next to him.

Living in Vanport, the girls were largely cut off from the outside world. They did not get a newspaper and had neither a radio nor a telephone. They knew what was going on in the wider world only through vague talk at work, and they kept

From left to right, Helen, Betty, Clarice (seated), and Marcelle
with Slim's car in front of their Vanport apartment, 1944.
Betty Anderson Bergman collection

in touch with their families through letters. Their own world,
however, was filled with new experiences. Soon after arriving,
Clarice broke the frames of her glasses, but she had no money
yet to have them fixed. Somewhere she had heard about a pawn
shop, a thing unknown in Stanley. She had a nice wristwatch
she had been given for graduation, so she and Betty went over
to Vancouver and found the shop, and the man inside agreed
to give her enough money to fix her glasses. "Please do *not* sell
my watch!" she told him. "I will be back for it for sure. Promise
that you will not sell my watch!" The man agreed, and when
Clarice got her next paycheck, she and Betty ran down and
reclaimed the watch.

Mostly the girls worked the day shift and relaxed on the

Clarice (with glasses) and her three friends pretending to
weld on the front porch of their Vanport apartment, 1943.
Betty Anderson Bergman collection

weekends. The group expanded when Betty's attractive sister Marcelle and another friend, Helen Domaskin, arrived
in Portland to become welders, too, and got an apartment
nearby. Smart dressers who were formerly secretaries, they
could make as much in their leathers each week as they earned
back home in a month. Being older (that is, in their twenties),
they were much more interested in opportunities for meeting
men. They took the younger girls to Jantzen Beach, a nearby
amusement park, and danced under the stars in a big open-air
pavilion to the music of Benny Goodman, Glenn Miller, and
other big names. But in the fluid atmosphere of wartime, one
didn't need to go far to meet good company—it could happen
just about anywhere. One day as Marcelle left to spend some

time by herself in Vancouver, Clarice and the other girls said, "Bring back some sailors!"

Marcelle went downtown, and as it happened, she did meet some sailors. They asked her, "Got any friends?"

"Matter of fact, I do," she replied.

At about 5:00 P.M., Helen looked out the window and saw Marcelle approaching with five sailors trailing behind. There was a scramble to get into the bathroom to comb hair and put on lipstick. Characteristically, Helen was the last one out: "I got the last sailor—not bad, but not the pick of the litter, either." They took the men over to a dance in Ogden Meadows and had a good time, but they never saw them again. They may have shipped out the next day.

The girls' encounters with sailors might sound risky today, but it was a more innocent time, a time when people were not so afraid. Aside from one dance where Clarice and Betty had to plan a furtive escape from a worrisome suitor by rendez-vousing in the ladies' room, the girls never felt threatened. "We could travel around anywhere, anytime and feel safe," recalled Helen. "Most of the people were either from work or in the service and were small-town people away from home during the war." They saw the sailors as nothing more sinister than young servicemen looking for girls—which was convenient, since they were girls looking for young servicemen. Clarice had several boyfriends during this time; none was serious or long-lasting, but she wrote to a few while they were away at war.

Wages, though, and not service or servicemen, were the main attraction that drew these young women to the hard work on the coast. As Clarice observed, "While everyone wanted to think people came for the patriotic war effort, and contributing to the war did make people feel good, in point of fact people were drawn mainly by the very good money they could make." Shipbuilding was one of the highest-paying industries in the country in both peace and war. In 1941, women in all manufacturing jobs made an average of around twenty dollars a week, while men earned about thirty-six. But Clarice made ninety cents an hour, or thirty-six dollars a week, as a female shipyard *trainee*. "How good a deal is that!" she thought. When she became a full triple-plate Navy welder, she earned sixty dollars a week, which she thought was a fortune: "When we got our checks, we were blown away!" The girls cashed their paychecks at the bank in the shipyard, faithfully arranging for the bank to send every other check to be deposited in their savings accounts at home. Betty's sister Dorothy worked in the bank in Stanley and told her, "Banker Nelson was so impressed to see you girls saving half your checks!" Working as a welder, Clarice saved enough money to pay for her first year at college.

When Evelyn graduated from high school in June 1944, she immediately left Stanley to join Clarice and Betty in their tiny Vanport apartment. She, too, intended to make her fortune in welding, but when she went to Kaiser to apply, she was disappointed. "Oh, you're just seventeen," they told her. "You can't get into the shipyard itself." She was devastated, but with typical pluck, she immediately started looking around for work outside the yard and found it in a tiny two-sided store, where one side sold ice cream cones and the other corn dogs. There was a window in the wall between the two halves, and at lunchtime the kid who worked the other side would poke his head through and offer to exchange a corn dog for a cone. She worked at the store for a few weeks. Henry Springan's son Arne happened to visit one weekend and snapped her picture.

But Evelyn wanted a better job, one more closely connected to the shipyards and with better pay. She noticed a number of buildings that sat right outside the shipyard entrance, arranged to create a kind of passageway leading into the yard itself. They contained shops, and like the yard, these shops were open all hours of the day and night. Workers passed by on their way into the yard and again on their way home. One of the buildings was a huge canteen or base exchange. Cavernous and drafty, it contained a cafeteria, a tobacco stand, an

Evelyn at her ice cream stand job in Portland, taken by Arne Springan, Summer 1944. *Richard Edwards collection*

ice cream fountain, and other counters that sold film, stamps, and other personal items. It was so big that it took fifty or sixty people to staff all the different stations. Places like this, because they paid less than the yards and had trouble keeping their help, always had jobs available. Evelyn applied for work there, and after giving her a quick interview, the manager said, "Well, go over there and get to work; can you start now?" He assigned her first to the tobacco counter, where yard workers would come up and request "Pack of Brown Mule, please," or "Give me some Camels." She didn't know much about tobacco but learned quickly. Then she rotated among the various counters, including the ice cream fountain, working wherever she was needed. When a shift ended in the yard, workers came streaming out in throngs; the building filled up, and everyone wanted to buy something quickly and be on their way home. Pretty soon the crowd would drain out, and it would be quiet again.

Evelyn rotated shifts, though she rarely got the night shift, perhaps because she was so young. Mainly she worked the swing shift (3:00 to 11:00 P.M.), which she loved. She had a boyfriend, Jack, who worked the same shift in the yard, and when he emerged a bit before midnight, they would go out, first to a triple-feature movie and then to breakfast. Such a routine was inconceivable in Stanley, and Evelyn found life around the yards to be wonderful fun.

She lived with Clarice and Betty, but because they worked different shifts, the two sisters didn't see much of each other. Typically Clarice left for work before Evelyn returned, so she didn't even know whether Evelyn had come in from the night or not. For both girls, it was a time of exhilarating freedom. One time Evelyn and Jack went out to hear Jack Teagarden, a renowned trombonist who later teamed up with Louis Armstrong. "I am really in the big-time now," she thought as she experienced the music. "I have arrived!" Although the war was terrible, Evelyn relished the strange, sweet quality that it lent to the times—a common purpose in the war effort, yes, but also a deeper sense that "we better do this now, who knows what's going to happen?" When Evelyn left Portland, Jack traveled east with her as far as Coeur d'Alene, where he got off the train. They parted, and she never saw or heard from him again.

By the end of the summer of 1944, the sisters' war was over. Clarice and Evelyn quit their jobs in mid-August and began their trek back east. Although they didn't think about it at the time, Clarice's departure from the shipyard and the departure of millions of other women from their wartime industrial jobs marked the end of an unusual time in American history. Clarice and Evelyn had other aspirations to pursue, but for most women workers, the end of the war also meant the end wartime opportunity. Women had proved that they were capable of doing "men's work." Yet that old idea, that in ordinary times women do not belong in shipyards any more than men belonged in foxholes, meant that with the war over, high-paying industrial jobs were once again reserved for men. It would be two full generations before the women's movement and changing laws opened them up again. Few women in the meantime would have the opportunities that Clarice, Betty, and other female welders had experienced in the Vancouver shipyard.

After a brief stay in Stanley, Evelyn and Clarice packed their belongings and set off for college. Few of the thirty students in each of Stanley's graduating classes went to college, and those who did mostly went nearby to Minot to attend the teachers' college or a business school. A few, like their friend Warren Flath, ventured the two hundred miles to Jamestown College, a familiar church school, or like Arne Springan, traveled across state to go to the state agricultural college in Fargo. No one thought of going all the way to the Twin Cities for college, but that's what Evelyn and Clarice did. Before her wartime adventure, Evelyn had taken English and home-economics courses in high school from Mary Lou Allison, and she thought it would be fun to be a teacher like her. That meant going to college, but she did not yet have any money to pay for it. Mary Lou's husband Clyde, the Presbyterian minister, suggested, "You can work in the summer and then go to Macalester. They'll help you out." Neither Evelyn nor Clarice knew anything about Macalester College in Saint Paul, but Reverend Allison's advice was so respected that neither of them ever considered going anywhere else. Clarice later said, "It's amazing. . . . Somebody makes a suggestion like Macalester, and [touching her temple] you think, 'Hmmm, maybe that would be a good idea.' You just sort of take a turn in the road, and that's what happens. Somebody says, 'That would be a good idea,' and if you take that chance, you do it, sometimes it's good, and that's your life."

When the girls arrived at the station in Saint Paul, they felt as green as could be despite their experiences in Portland. They took a cab to the college, where the driver dropped them off on the street with their trunk. They saw the whole college ahead of them but had no clue where to go. It was early evening, and darkness was coming on. Then a young guy with a withered arm came over and greeted them.

"You looking for a dorm?"

"Yes, we haven't ever been here before."

"Ah, I can help you; come on, what dorm are you going to?"

He grabbed their trunk and threw their loose items over his arm, and in so doing, he welcomed my family to college. "It couldn't have been nicer," Evelyn said later. "I remember him so clearly, I don't know his name, and I thought so many times, I wish I had kept in touch, because I'd love to write him a letter and tell him, 'You don't know how much that meant to us!'" Using their savings from the shipyard, Clarice and Evelyn became the first members of our family to enter college—and having started college, they were determined to finish it. "We just thought, we *will* graduate," Evelyn said. And they did, in the class of 1948.

Two sisters, naïve girls from a little Dakota wheat town who lived at a time when most girls saw no reason to go beyond high school, had pluck. They insisted on bigger opportunities. Their parents and other community members encouraged them, but ultimately they had to make their opportunities themselves. Their pluck took them on an adventure halfway across the continent to do work that had recently been unimaginable in order to achieve their dream. Sixty years later, Evelyn said, "When I look back on it, we were gutsy in a way, but it wasn't considered that at the time.

Because I think all young people felt that way, I imagine they still do today. . . . We were absolutely convinced that [life] was going to get better, we had some [of] what we'd certainly have to say were some pretty difficult times, but it never seemed like we were depressed. . . . We never had times where we just couldn't see our way out of something."

When Clarice and Evelyn left Stanley to work in Portland and then went on to college, they set my brothers and me on a path that changed our lives. They had pluck, a spirit of adventure, optimism, and determination. Clarice described their attitude best: "People always thought things were going to get much better. They were never afraid that 'I can't put my child through college.' . . . We were always advancing, your second apartment was a little bit better than your first apartment." Glancing around her four-bedroom northern Virginia house as she spoke, she recalled how she and her husband had felt when they moved into their second apartment: "We moved just before [our fourth child] Nancy was born, and we [now] had *two* bedrooms, and it was fabulous!"

COMMITMENT

· · · · · · · · · · ·

CARELESS LOVE, ENDLESS LOVE

· · · · · · · · · · ·

She met him in Missoula on January 4, 1966. She had driven down from Polson and went straight out to the airport to meet the flight from Bismarck.

They hadn't seen each other in nearly thirty years. Irene, my mother Winnie's baby sister, was forty-six and recently widowed. Arne Springan was fifteen months older and still married—still very unhappily married. Neither knew what to expect. For the last three months they had written to each other, talked secretly on the telephone, and even exchanged tapes that they had recorded for each other. Still, this reunion was the big test, and both were nervous. In his last letter before the meeting, he had written, "I don't know what I'm going to do when I come. Don't be surprised at what I might do!"

She stood in the terminal waiting for him; he was the last one off the plane. There was a last moment of joyous disbelief. He had dreamed about this meeting and patiently worked toward it for years. Now it had come to pass. He picked his way down the airplane's steps and strode through the doors to where she stood waiting.

"Hi," he said.

"Seems like only yesterday," she responded. "Oh it's so good to see you, I can't believe this is happening."

"Oh my god, it is, though."

For both of them, insecurity and doubt about what the decades might have done to each other and to themselves quickly faded. The twenty-nine-year gap in their relationship evaporated, and they talked easily, as though they had last seen each other only yesterday.

They got into Irene's car, Arne slipping into the driver's seat. Then they drove around, just to be together and talk.

"How long can you stay? When do you have to go back?" she asked.

"I only have three days. Wish it were more."

They didn't yet dare go too deeply into their emotions. Arne's natural reticence and Irene's ambivalence about meeting a married man limited their talk to superficial topics. She asked whether he had heard from anyone else in their graduating class, and he reported the news.

Still, both of them recognized that something big was happening. This fourth day of the month promised to be an important one in their lives. They knew that it was the beginning of a second chance, an opportunity to reverse the careless and disastrous decisions they had made in their youth, choices that had separated them and left them both desperately unhappy. Yet anxiety tempered their joy, since their reuniting threatened to hurt others whom they loved and represented a violation of Arne's commitment to Betty, his wife.

Folks in Old Stanley believed that a person's commitment held force through good times and bad. When a man or a woman committed to do something in business or the community or in personal life, he or she was expected to live up to his or her word. A contract was not necessary, because his or her word had been given. Honoring commitments had special force in marriage. Neither changed circumstances and fortunes nor altered affections lessened the obligations of commitment, and couples tended to stay married until one or the other died. For some this belief was religious. The Catholic church was the most explicit in prohibiting divorce, but Lutherans and others frowned on it, as well. Even among the non-religious, divorce had an aroma of shame and failure. For years after my grandparents, Rosabell and Webster Edwards, were divorced, my father could not speak of it; he acknowledged only that they lived separately. When Irene and Arne met in Missoula, Irene's husband, Al, was dead, but Arne's wife, Betty, was alive in their Bismarck home. His breach of his marital commitment made Irene uneasy. She could not necessarily let herself believe that he was as unhappy at home as he said he was, and she disapproved of her own role in his failure. "Fast coming is my day of punishment," she told herself. Arne, too, agonized over his actions, both past and future. They would "have to be turned over, upside down and sideways in my mind," he said. "It is going to require decisions that to me are going to be terrifically difficult to make."

On the other hand, there was also the commitment of their love for each other, an intense, enduring, exhilarating, lifelong love. How did that enter the calculus of obligations? How did it alter the meaning of commitment? Or was it simply irrelevant? What claims did their own happiness have when weighed against the damage they might do to others? Both Irene and Arne had dreamed of undoing their fateful early decisions, but neither one ever really expected it to happen. It is rare to get a do-over in life, yet now, perhaps, they had the chance. But at what cost? In their torment to find what was right, we see the strength of Old Stanley's virtue of commitment.

Irene told me her story when she was ninety, seven years after Arne died. Until then, I had been only dimly aware of the arc of her life. As she recounted her life and I began to realize its full drama, I wished that I could have heard Arne's voice, too. I had hardly known him. A private, laconic man, he would have been reluctant to share his intimate memories with me. But Irene told me that shortly before his death, he had written his version of their story. She went to retrieve it from her collection of old letters, tapes, photos, and notebooks. She returned with a twenty-six-page document, inexpertly typed on a manual typewriter and corrected in Arne's own handwriting. Here was his story.

Irene and Arne had been best friends since second grade.

Irene at ten years old, in a photo taken by Arne in 1929. *Richard Edwards collection*

He lived on the southwest side of Stanley, she on the northeast, and they met at school. He was aware of her before she was conscious of him. In fourth grade, at the school's spring picnic, Arne said, "Hey, Irene! Let me take your picture!" and he snapped his first photograph of her. She was ten, and he was already smitten. He couldn't get his eyes off her, or his mind. At parties, at picnics, or playing in the schoolyard, he maneuvered to be close to her. He held himself in the background until he could move unobtrusively into a position where he would be paired with her. He wanted to be in a place where kids would, as kids do, tease him, "She's your girlfriend!"

By the sixth and seventh grades they became part of a group of kids more or less their own age who played together. They skated and went to church socials and had parties in their homes where they played spin the bottle. They were always with the group, but also slightly apart, together. Their relationship deepened when they entered Stanley's small high school in 1932. Arne especially was absorbed with Irene, always wondering what she was doing, where she was going, what she was thinking. At assembly during freshman year, his assigned seat was third from the front on one aisle, while she sat far away in the third seat from the back on the second aisle. They exchanged notes, passing them through a half-dozen kids whom they trusted not to open them. Arne: "I think I can get the car tonight.... I'll be at the store at seven.... Will you be home?" Irene: "Of course!"

Arne was growing up to be tall, thin, lithe, athletic, and strikingly handsome. Although he was quiet and reserved, he became something of big man around the high school. His sophomore year he tried out for the football and basketball teams and made both. His new prominence cast the first shadow on their relationship, as girls in the upper grades began to take notice of him. Although Arne stayed true, he often found himself at parties with them while Irene, as a lowly underclass girl, was excluded. Jealous and angry, she quarreled with him, breaking things off and patching them up again as young lovers do. One time, after a tiff, Arne sent his younger brother Paul over to Irene's house to find out if he was still welcome. As usual, she relented.

Irene was a contrast to Arne. She was much shorter, with a pretty, round face and a fuller body. She liked hats, much in fashion at the time, and she was rarely seen without a beret. She was lively, full of teases and jokes and personality and motion. In photographs, Arne was always the sober, serious one, while Irene was the creature of many poses, the girlish, grinning one.

Most kids and their parents had little money during the mid-1930s, but Arne had two jobs, delivering for Hansen's Grocery and working in his father's furniture store and mortuary. He began driving at ten or eleven, and by high school he was allowed to take the family car out by himself. He and Irene would drive out of town for picnics or to walk along the Knife River. Practically every weekend he ate Sunday dinner at Irene's house, and during the week he frequently came over to study with her.

Summers were magical times for them, and they spent the seemingly endless, lazy hours together. Although Arne continued to work, his hours were flexible and he could use the family car whenever he wanted. Early in the morning he would drive over to her house, go around to the back, and knock on her bedroom window: "Ready to go?" She came out and they went to play tennis before the day got hot. Later they would drive out of town to shoot gophers or rabbits. Or they might drive over to Chocolate Drop, an unusual hill that rises a couple of thousand feet above the prairie out near Ralph Hagey's homestead southeast of town, where they would hike and have a picnic. They might go to the Sweet Shoppe or play golf. Other times they would drive down along the Missouri River at Sanish and stop at a truck garden to buy watermelons.

When school started their senior year, they spent almost every evening together. On weekends, there were ballgames and dances; on weeknights, they studied at the kitchen table at Irene's house. Her parents, Frankie and Bill, would sit in the front room, and occasionally Frankie would wander out to the kitchen to visit with them: "Are you about through studying? You hadn't better stay up too late." Later still, she would hint more strongly, "Irene, it's getting late," so they would go out back, leaning against the porch railing to say long goodbyes. "I can still feel the imprint of that railing where we sat," Arne remembered later. "So many hours of conversation when not a word was said."

Irene's and Arne's relationship was close, but it was not fully sexual. Staying on the right side of that line was more frustrating for Arne than Irene, but he agreed that crossing it would be "doing things that could not be done," and that if they had lost their control, "we could never have lived with ourselves again, with each other or with ourselves." Irene's parents had such confidence in Arne that they were unconcerned about the temptations the young couple faced—a remarkable attitude, considering that Winnie had become pregnant at seventeen.

Irene and Arne knew that they would be married; he never

proposed because it was unnecessary. "We will have two children," she told him, and she already had names picked out. Irene didn't expect to go to college—neither she nor her parents had the money for it—or to have a career. She idolized Winnie's family, whom she saw as being happy, and wanted to replicate her sister's success. Arne did not know what he wanted to do, except that he was certain he didn't want to be a mortician and wasn't much interested in selling furniture, either. He figured he might find out at college.

Their senior year ended as gloriously as it had begun. They both had roles in the school play, *Room for Ten*. At graduation, the local American Legion post gave out its annual awards to the male and female graduates who were the "best all-around students," based on grades, extracurricular activities, and personal character. The awards went to Arne and Irene.

That summer—the summer of 1936—was an enchanting time. Arne had now grown into his full physical beauty, stunning in his tennis whites, his blond hair, and his tall, straight posture. Irene snapped pictures of him in a field where they stopped for a picnic or leaning against his car with his hat at a rakish slant. Irene, too, seemed to glow that summer, her pictures showing a smile and a tilt of her head that intimated how vibrant and full of hope she was. But as summer wore on, they began to confront the change coming to their lives. They had no plan, but Arne decided to attend the state agricultural college in Fargo. He was not interested in agriculture, but it was a place to start. He went to see about getting admitted and found a job for the fall with help from an uncle from Detroit. Irene decided to go to Minot for several months to study for a "commercial," or secretarial, certificate. They spent some sad days, sad evenings, sad weekends thinking about their coming separation; they had not been apart for more than a couple of days since early in grade school.

Finally the day came for Arne to depart. He picked Irene up early that morning so they could say goodbye, then he boarded his train for Fargo—the "most horrible train ride I believe I

Arne and Irene at the
Sanish bridge, Summer 1936.
Richard Edwards collection

have ever had." Later, Irene left for Minot and her commercial course. Arne grappled with loneliness that fall. Irene wrote him a letter every two or three days, but he agonized on the days without letters. She seemed to get along better with her studies but also missed him.

They both returned for Christmas, but now Arne was burdened with some heavy information. While in Minot, Irene had gone out with a boy who was also taking the business course. They went to the movies a few times with another couple. It had meant nothing to Irene, and she had never mentioned it to Arne. But he had heard about it from a high-school classmate attending Minot State College, who said that Irene was running around with another guy. Arne was deeply hurt, but he never confronted Irene about it because, he told himself, "Well, you can't complain, she has to do something but go to school."

Irene finished her course in April. On returning, she learned that her parents had made a huge decision. Not prospering in Stanley and looking for a new start, they were leaving town and moving to Polson, Montana, up in the mountains. Irene, just eighteen, with no place to live, no job, and no reason to stay, went with them.

When Arne got back to Stanley, he found "the most nothingness I could ever imagine." Stanley was empty for him. He thought about going out to see Irene, but the obstacles seemed insurmountable, the trip only a fantasy. He cooked up an excuse to see Roy, hoping to get some news of Irene without directly asking. Roy was in his workshop out in our garage, finishing his new house trailer. He mentioned that they were driving to Polson to try it out and asked if Arne would like to go along. Arne's mind exploded with possibilities. He quickly secured his parents' permission, and two days later he was in Polson.

Irene worked for an attorney during the day, and in the evenings she worked from six to ten at the Bon Ton Cafe, her parents' confectionery store. Arne stayed with Irene and her parents at their house; all the bedrooms were already occupied, so he slept on a cot in the wide hallway. He found work as a painter. He hated the work, but to him it was "the most wonderful job I ever had" because it allowed him to stay longer near Irene. Their summer was filled with days and evenings on Flathead Lake in a borrowed boat, swimming, and taking drives up into the high mountains to picnic. They spent hours together, just the two of them, and the emotional pitch of their lives was as intense as it had been during the golden summer of 1936.

Early mornings, after others had gone to work, they were alone in the house. One morning when it was still dark, Arne came into Irene's bedroom and climbed in with her, Irene in her red pajamas with bell-bottom trousers and a striped blouse. "We shouldn't be doing this," she protested, unconvincingly. They were both feeling passionate, but again they controlled their impulses, snuggled down, and went back to sleep. Much later Arne wrote, "We could have ruined our lives as many I see now have been ruined and live with memories they can't forget but God how they wished [they could]!" The morning was one that Arne would replay in his mind for years, questioning whether things could have been different had they made the other choice.

Toward the end of September Arne received a letter from his dad telling him, "It's time you came back home." Sadly, dutifully, he packed up, said goodbye to Irene, and left. And that, with one painful exception, was the last time he saw her for twenty-nine years.

Lacking the money to go back to college that year, Arne got a job in the Dakota Drug on Main Street. He and Irene wrote to each other almost every day. Soon, however, he was writing more often than she. Irene continued to work her two jobs in Polson and went out occasionally with some friends. Then one day Al Pronovost walked into the Bon Ton, and they chatted while he drank a few Cokes. He came back to see Irene several times. Al was handsome and charming, with dark American Indian looks. He worked down at the dam, a good job, which meant he had a car and some money to spend. Al liked to drink and smoke and play cards for money; he was fascinating. He and Irene started going out in September.

Irene and Al were married on December 24, 1937. She sent Arne a letter beforehand to let him know. She knew it would be a cruel letter. Arne received it on a leaden day with no sun and no snow to brighten Stanley. He was devastated. He doubted himself, wondering why he was not strong enough to go out to Polson and win Irene back. And he immediately forgave her. In his grief, he wrote to her: "Irene, if you ever need help, don't hesitate to call."

He tried to convince himself to move on. "Okay, it's past," he told himself. "Let's forget about it. Live your own life." It was useless. "These are not solutions to me," he thought, "because . . . it wasn't only love, Irene and I were one. It could be no other way."

Days after she and Al were married, Irene knew she had made a careless, disastrous mistake. Instead of the carefree life she had enjoyed with Arne, one of innocent fun, hiking and picnicking and playing tennis and swimming, she found herself in a different situation. She later convinced herself that she had not seen the drinking and gambling sides of Al's personality during their courtship, but perhaps it was exactly his "bad boy" image that had attracted her in the first place. Once the thrill of the early attraction faded, she discovered that they had almost nothing in common. Neither of them was happy, and Irene retreated into her work. Al's family were devout Catholics, so Irene began studying to be converted, but she abandoned hopes of having a happy little family like Winnie. Through rashness, she realized, she had lost everything.

Irene, never self-reflective, said that her unhappy marriage was simply a mistake, and that was that. Some family lore said she did it on an inebriated dare. Arne blamed himself, believing that over many years he and Irene had developed a fever of desire and emotion, perhaps reaching its agitated peak that early morning when he climbed into her bed. And then—nothing happened. Nothing ensued, nothing was consummated. And when he left Polson, Arne surmised, Irene "went over-board." He thought it was his fault that Irene had made such a ruinous decision.

Irene and Arne saw each other again briefly when she visited Winnie and Roy in Stanley the next summer. One day she and her friends Marguerite and Maxine went to the Dakota Drug for a Coke—hardly, one thinks, an innocent destination. Arne was working the fountain. After the girls left and started walking home, Arne drove up and asked if they wanted a ride. He took Marguerite and Maxine home first and then drove Irene to Winnie's house. They talked for a few minutes. Arne pulled out his billfold, extracted a picture, and showed it to

Irene. It was a picture of his new girlfriend. "Life's not that bad for me," Arne said. Irene felt a knife twisting in her back. When she left Stanley, Arne was there at the depot, but he stayed hidden, not daring to let her catch a glimpse of him. And so began a long, long silence.

Irene and Al tried to have children, but couldn't. The doctors said that there was nothing physically wrong with either of them—they were simply incompatible. They adopted a beautiful little girl, Yvette, who like Al, was part Indian. With Yvette's arrival, Irene found more fulfilment, and despite their incompatibility, the couple stayed together and tried to find common ground. They became regular churchgoers, and Irene served as secretary to the local priest for many years. They owned a boat and spent time on Flathead Lake. Despite their efforts, they drifted apart, but as devout Catholics, divorce was out of the question. When World War II came, Al enlisted and spent much of the war abroad. After he came back from the war he seemed more depressed, and though he never abused Irene, his self-destructive behavior—heavy smoking, heavy drinking—increased. Yvette grew up, and the years passed. Some years were better, some worse. After his first heart attack in 1955, Al suffered a series of heart attacks, and the last one claimed his life. It was June 1962. He and Irene had been married for twenty-four years.

In the meantime, Arne's life had not stood still. He spent four years studying optometry in Chicago. He worked hard during the week and spent the weekends trying to forget Irene. He began to correspond with Betty, a young woman in Stanley who was on the rebound from a breakup. She and Arne acted like a couple, and although Irene was still on Arne's mind, somehow they agreed to get married.

At midnight on December 21, 1941, fourteen days after the attack on Pearl Harbor, Arne arrived in Stanley for his Christmas vacation. The next day he was to be married. He later described thinking that "I might as well go this way; the path of least resistance—because I have no other way to go." The next day, his wedding day, Arne searched through the belongings he had stored away in his room at his parents' house before leaving for Chicago, looking for his album of Irene pictures. At 6:30 that evening he went over to the church, and he and Betty were married. The newlyweds left for their honeymoon in Minot that same evening. It was snowing. Thinking of Irene, Arne was miserable, "the most miserable I have ever [been] in my life, so help me God!"

After he finished his schooling, Arne joined the service. In 1944 the army sent him to study Russian in Corvallis, Oregon, and he learned that Irene's parents and nieces were living nearby in Vanport, working in the shipyards. He decided to visit them, "with the fear or the hope, whatever it may have been" of seeing Irene, too. Arne saw Evelyn at her ice cream stand and snapped a photo, but there was no Irene. Out of consideration for his feelings, neither Frankie nor Bill nor anyone else ever mentioned her. Arne returned to Corvallis and was shipped overseas. At the end of the war, he was despondent about returning, "because I already knew I had nothing to come back to except one thing. The love for . . . my child would provide me with existence only and nothing else."

He and Betty had had one daughter before he went to war. "They say children cement things," Arne said. "We had two more. All they did was cement the ties between myself and them with no help at all to the other situation." Arne was close to all of his children, but otherwise he fled his home. He became a joiner, searching desperately for activities that could occupy the long, unhappy hours when he was not working. For a while it was the Boy Scouts with his son, then the Masons and the Shriners—anything but staying in the house.

Arne subscribed to the Stanley paper, the *Promoter*, to get news of Irene, but he saw her name only a couple of times during those years. He didn't know where she was or what she did. Every year he waited impatiently for Winnie's Christmas card with the family news, hoping she'd slip up and give him some news about Irene. Thinking that she was being kind,

Winnie made sure she never did. The years wore on. In 1956 Roy and Winnie stopped in at Arne's office in Bismarck during their move to Virginia. Arne desperately wanted to ask about Irene but didn't dare, and once again, for fear of opening old wounds, they carefully avoided mentioning her.

Arne took it hard when his first child, Gail, left home in 1961. When Mark left a few years later, Arne realized how empty his home would be without his children. He went on business trips whenever he could and began to spend most of his time at work. He left home around 8:00 in the morning and often stayed until 2:00 or 3:00 A.M., not because his growing business needed so much attention, but because he couldn't face going home. He was desperately lonely and unhappy.

When he heard that Al had died, the news stunned him. Arne decided right then that regardless of what she said, he would work to meet Irene. Yet he hesitated to act because he was afraid that she would treat him as just another old-time pal; that they would only have a cup of coffee and say goodbye and good luck. He thought many times about calling her but was paralyzed by fear. As he told her later, "Compassion spilled out of me as it never has before, but something else welled up in my mind and I guess fear came. Could this dream of so many years somehow yet come true?" He was harsh in his self-doubt: "Am I destined to be denied something that I have wanted so long because of something I don't dare do?"

He had made a habit of browsing in shops looking for the perfect card to send Irene, a card that would seem to be its own excuse. One day in July 1964, more than two years after Al's death, Arne was killing time after a professional conference in Indianapolis when he saw a card with a duffer on the front. It said, *"Golf is a way of life."* Opening the card, he read, *"Take it from one who's played a-round!"* Remembering that they used to play many rounds together, and perhaps liking its naughty suggestion, as well, Arne decided to send it. He tried to write some nonchalant news about himself, what he was doing, and his kids. Of course, he carefully avoided any hint that he was lonely. After the trivial family news, he casually mentioned, "Thought a couple of times I was going to get to Missoula for some meetings, but they never materialized. If so I'd thought of renting a car and running up to visit you all. Maybe someday." He concluded, "Don't know if this will reach you or not but will send it on its way. Arne." Then he sealed it and put it in the mail.

"If I could have reached in the mailbox after I had dropped it in I probably would have gotten it back," Arne said. "But it was gone." Then he waited, desperately hoping that this nonchalant card would be answered quickly.

Irene remembers the day she reached into her post-office box and withdrew the envelope addressed in that instantly familiar handwriting. Her mind flooded with memories. She opened the envelope and read Arne's long, dull message. She was crestfallen. She put the card away, up on a top shelf. Several times over the next few months she took it down, read it again, and replaced it, vexed each time that it was so impersonal.

Meanwhile, Arne waited for a reply. There was none. He waited. He waited some more. No word. He waited a year and four months, until November 1965. On the day before Thanksgiving, he went to the post office, "and there, cattiwampus, face up, was an envelope that almost made my heart stop. I saw it; I knew it; I recognized it."

"There's no special reason for this card," the front of the card said. Inside, as in the card he had sent, the preprinted message was suggestive of deeper feelings but did not venture too far: *"It's just that sometimes I get this uncontrollable need to communicate with a friend."*

"Have intended to write you a note for a long time and let you know that I appreciated the card that you sent some time ago," Irene wrote. "It was so nice to hear from you! Even had toyed with the idea of calling. It would have been real fun to have talked with you. Sorry that none of your trips in this direction have materialized. Think it would have been interesting to see how you have aged. Of course, I haven't, you know." She, too, added some trivial news and concluded, "May take a long weekend this winter and fly down to Las Vegas. Have a friend down there who is trying to promote something. I'm being a little cautious tho; I made a mistake once! Hope to hear from you again! Irene."

Arne was as deflated by Irene's nonchalance as she had been by his. "It made my heart sink just a little bit. Not just a little bit, I guess, it sank right down to the bottom." Yet he kept coming back to her comment about having made a mistake once. Was it just an insignificant aside? Or did it mean . . . ? "The dream re-opened," he said. There was a chance.

He decided to call her. He found the number for the bank where she worked, checked and rechecked the difference in time zones, and went home, where he would be alone in the middle of the day. He dialed the number.

"Hello, this is Arne."

"What?"

"This is Arne. I got this number to call you, but I don't want to talk now, I just want to get your number at home so I can call you tonight."

She gave him the number.

"I'll call you at eight. Okay?"

She agreed and hung up.

Irene's heart was beating like crazy. The day dragged endlessly for her, and she couldn't concentrate. When he called that evening, she was just taking her Thanksgiving pies out of the oven. The long years, the long decades of not talking, not hearing, not knowing, were over.

They talked for over an hour. On the surface, the conversation was nonchalant. They covered what their children were doing and what Arne and his brother Paul were up to in their optometry business. Nothing more was spoken aloud, but beneath the surface, sparks were flying as their minds raced with anticipation, regrets, alarm, hope. Arne sensed—perhaps

Irene and Arne at the Missoula airport a few months before
they were married, 1968. *Richard Edwards collection*

wishfully—that things really hadn't changed much. He asked,
"Can I write to you?" and she agreed. He posted a letter that
night, including pictures of himself and his kids, and so began
a frenetic exchange of letters. Hers, of course, were addressed
to his office, so Betty would not find out.

Right away Arne began thinking about a meeting. Would
Irene come? He thought she wanted to, but would she dare?
Despite the doubt, he was determined to press ahead. For her
part, Irene worried whether meeting him was the right thing
to do. Moreover, she had seen cases in her work that showed
her how unhappy such matches could be. Now their roles were
reversed; now he was married and she was not, and she worried that she could get hurt. He told her how unhappy he was,
but she said, "You're only unhappy because I've now become
available; I'm the cause." Still he persisted and finally wore her
reluctance down. He said he had seen their meeting coming
since that first phone call, and she agreed. Arne responded,
"God, Irene, I want to [meet] and I'm going to turn it into a
life of joy."

About six weeks after that first phone call, Irene met Arne

at the Missoula airport. Driving around in her car, they reconnected and talked as in the old days. They went out for dinner. Still, although each was swelling with emotion, they didn't allow themselves to reveal much. There were no expressions of love that evening because they were still feeling each other out, wary of being hurt again. They ended the evening at a hotel, in separate rooms.

After three days, their visit was over, but it unleashed a blizzard of letters. Arne wrote virtually every day, Irene two or three times a week. She worried that her letters might go astray in Bismarck, but Arne assured her that he always picked up his office's mail. Then he suggested that they send each other tape recordings; he told Irene what type of recorder to buy, and they exchanged tapes on a regular basis. They found these tapes to be even more personal and expressive than their letters. Arne's, in particular, were filled with struggles, emotions, and hopes.

They discussed whether to meet again. Irene protested, "Isn't what we're doing wrong?" Arne felt that he couldn't wait to see Irene until after he arranged matters with Betty, because he dreaded losing her again. Now he knew he had to get a divorce, but the thought of hurting his family tortured him. "It is going to require decisions that to me are going to be terrifically difficult to make," he told Irene. "I think in this with me you will agree. You hope that the decisions are not too great nor that they shall bother me too much. They will be some of the greatest, Irene, I have ever had to make. Bother me, oh yes, they'll bother me. . . . I can't say that I'm brave in carrying these loads but I have tried to carry them as best I can. My decisions, Irene, are going to be mainly concerned with how this can be done and cause the least, if you can say there will be anything like the least, amount of tears and heartbreak for those, yes all of them, those I love."

He went on, "You mentioned in your last phone call that . . . when I do [these things] I will have to have made up my mind that I do them without regret. I don't think you quite meant that, Irene. . . . I do not believe you would be you in asking me to do these things and have no regrets. You can't have lived this life, and the most important part of it is the family part, for all these thirty years, without experiencing regret. A person wouldn't be human at all if they did."

Arne felt special anguish over what he was doing to Betty: "I have real trepidation in my heart, Irene, in that I have to hurt somebody that I can sincerely say I have much respect and admiration for and a certain type of love; I think it borders a little bit on mother love only she happens to be the mother of my children. . . . I don't want to hurt her and this is what I am afraid I am going to have to do. This bothers me greatly. I hurt her first by letting her become my wife when I knew, probably not as well as I know now . . . that there would always be someone else who had my heart. There was none left for anyone else. Think she'll understand that? No, no one can but you and I. This is going to be a real hard thing, Irene."

In his pain, Arne needed someone else to talk to, and he found a confidante in his brother Paul's wife, Mae, who was a mutual friend of his and Betty's. When Paul went away for some meetings, Arne took the opportunity to talk to her. They chatted for an hour and a half before he worked up the courage to raise the issue that was really on his mind. Mae was not surprised at all, and they talked long into the night. At one point Mae said, "You know, your parents have worried for a long time about what was wrong with your marriage. Several times Henry asked me, 'Is anything wrong over there? There seems to be something missing. There's no family life there.'" Arne found it startling to hear that others had guessed about his marital struggles. From then on Arne had an ally in Mae.

Arne and Irene met in Missoula five more times and continued to exchange letters and tapes. During one of their meetings, Irene spent the whole weekend sick with the flu, and Arne plied her with tea and bouillon. Other times they would buy some bread and cheese at the grocery and go hiking up Lolo Creek, or have a picnic at an old logging camp, or trek down to see friends on the Thompson River. Each time, they grew closer and closer. Irene was still bothered that these meetings were not right, but Arne was insistent: "I just have to have another meeting to see you again; please do it!" At their last meeting, they became intimate. "I've been married to you all my life," Arne told her, and Irene agreed.

Despite his concern about hurting those he loved, Arne pushed forward with the divorce. Betty was shocked and unhappy, and the proceedings became long, bitter, and contentious. The divorce was finally granted in October 1968, in Minot, on grounds of incompatibility. Arne lost all of his assets except for his business in the settlement, so he needed to start over. But after a ninety-day waiting period, he was free to marry Irene, who had been skeptical that the divorce would ever happen. She was finally satisfied when she saw the decree.

Arne and Irene were married on January 7, 1969, at a chapel in Coeur d'Alene. Irene was forty-nine and Arne fifty-one. After honeymooning on the West Coast, they returned to Polson, loaded up a small U-Haul with her belongings, and drove back to Bismarck. Arne had found a new apartment for them before leaving, and they moved in as a married couple.

Irene feared an awkward life in Bismarck. Arne's daughter Dawn, a high school senior there, was perhaps the most vulnerable person in her parents' breakup, yet from the beginning, she was also the most supportive of Irene. Arne's friends and family were supportive, as well. Irene had worried especially about his brother Paul, fearing that he might still resent her for having hurt Arne all those years ago. But in 1967, when Arne told Paul that he had seen Irene again, Paul's reaction had been, "Really? That's wonderful!" Irene's fears proved happily unfounded.

Both of them prospered in their new life, Arne in his optometry business and Irene as a legal secretary for the North Dakota Supreme Court. Three years later they bought land

Arne and Irene in their kitchen, early 1990s. Arne's apron says, "You make my bean sprout." *Richard Edwards collection*

along the Missouri River six miles south of Bismarck, and together they designed and built their house, a big, comfortable dwelling with a lawn running down to the river. They were married for thirty-three years, living in palpable and delicious marital harmony. After all those years apart, they delighted in each other's company.

Arne remained the reserved, slightly shy, slow-moving man with a twinkle in his eye. All the Springans were known for being slow-moving—Arne's dad, Henry, was affectionately called "Flash." Irene was (and is) the quick-moving, funny, gregarious live wire that she had been when Arne first met her in second grade. Sometimes, when Arne was especially slow, Irene would work herself into an impatient tizzy. One time she asked him, "When I get like that, why don't you just shake me and say, 'Straighten up!'?" Arne simply replied, "Because I'm afraid I'll lose you again."

Even as they rejoiced for the couple's long-delayed happiness, there were those in Old Stanley, including some members of my own family, who considered it a shame that such happiness could be achieved only through the breaking of Arne's commitment to Betty. Arne and Irene themselves knew the strength of the virtue of commitment and the importance of the bonds it creates; what else could have made them hesitate to revive a relationship that was so clearly meant to be? In the end, the lifelong love that they shared also defines commitment, commitment of a subtler and perhaps more profound kind. In their hearts, as they agreed, the two had been married all their lives.

It was on January 4, 1966, that Arne picked his way down the steps of the airplane in Missoula to see Irene again after all those missing years. Afterwards, on the fourth of every month, Arne gave Irene a single red rose. Initially, when he was living in Bismarck and she in Polson, he had to arrange for a florist to deliver her rose at her workplace, where she took a lot of good-natured abuse from her co-workers. Later, when they lived together in Bismarck, Arne made the delivery himself.

When they were courting the second time, Arne wrote to Irene, "I don't think even though this story is told it can be understood by anyone but us." But perhaps he was wrong: great love is understood universally. Arne died in 2002, committed to his endless love, Irene.

DAUNTLESS OPTIMISM
· · · · · · · · · · ·
DARE TO BE A DANIEL
· · · · · · · · · · ·

Built in 1928, Stanley's graceful little Presbyterian church is admired as a gem of Tudor prairie architecture. Its "beautiful stained glass windows, . . . elegant supports, and grand openness" are said to "represent the dignity and beauty appreciated by its Presbyterian founders." Alas, architecture is not everything. On Sundays, while cars filled the parking lots of Stanley's Lutheran and Catholic churches and spilled out onto nearby streets, our church struggled to attract enough congregants to support itself.

It was hard to retain a minister on the meager pay that our church could offer. Clyde Allison, the minister in the early 1940s who pointed my sisters toward Macalester College, was a brilliant man, thoughtful and bookish, and also a nice guy who paid attention to people. He and his wife were products of elite eastern universities and eastern culture, but he felt his upbringing to be a burden, a result of too much immersion in academia. He wanted to meet real people in small rural communities, so he accepted appointment to Stanley's small church. Despite his sincerity, he was a misfit for Stanley, a fish out of water. People liked him because he was such a pleasant, genuine person who worked hard at his ministry, but after a few years he too left town, to be succeeded by yet another new face. The church's membership finally petered out entirely in the 1990s and the church closed, but the building was saved and repurposed as a community cultural venue called the Sybil Center.

The church depended mostly on volunteers for its upkeep. One time, my father Roy and some others were recruited to repaint the sanctuary. Although he was never a churchgoer and disliked painting, Roy was always willing to help out, as he did every Easter Sunday when he willingly joined the cook staff to make pancakes in the basement. On this occasion, the men were on scaffolding to paint the high ceiling, and Roy fell off. He badly hurt his back and spent several months at home in bed. He suffered no permanent injury, but the accident confirmed his belief that church was no place for him.

Through all the years, even as the church struggled to keep a minister and pay for heat and maintenance, there was one constant: the choir, led by the indomitable Isabel Flath. She had an imposing presence, tall, with perfect posture and a somewhat regal air. Her choir typically had ten or a dozen members and depended on the participation of high school students. Before he graduated in 1951, my brother Jack was one of three guys about his age who formed the tenor and baritone sections. One of them, Bob Jensen, could really sing; the second was okay; Jack's voice, according to his self-appraisal, was restricted to about half an octave, although he maintains that it was beautiful within its four-note range. As with many of her singers, Isabel had known Jack from a young age. In fact, when he was about three, she had recruited him to sing a solo rendition of "Away in a Manger," which he did without any ill effect. Isabel believed that everyone could become a creditable singer, even those who were confident that they had little talent. Some of them, to be sure, needed more help than others. In Jack's case, Isabel would sometimes rewrite his part of each hymn, giving him his own sheet of music with notes that stayed within his restricted range.

Irene, too, sang in Isabel's choir when she was in school, as did my sisters Clarice and Evelyn after her. In a small congregation, Isabel had to assume that everyone could do everything and admitted all who were willing to be part of the choir. "Poor Isabel," recalled Irene. "Always trying to have a good choir but it's kinda hard when there's no talent." Although an excellent and demanding musician herself, Isabel rarely criticized her singers, instead encouraging them to a better performance. She treated being in the choir and especially the weekly practices as something serious: you came on time, and you didn't fool around. Her professional attitude made the rehearsals seem important and worthwhile. Only infrequently did she become irritated, and then her face would redden and her singers knew that they had made a big mistake. At one practice she became so frustrated that, forgetting that she was holding her pince-nez, she slammed her hand down on the balustrade, breaking the glasses. The choir was first shocked and then—quietly—amused, but Isabel quickly recovered her composure and continued the rehearsal uninterrupted.

Isabel's determination that we Presbyterians would have a spirited choir was an example of the dauntless optimism that Old Stanley folks so much admired. Her attitude neither ignored reality—she knew the shortcomings of her talent pool—nor was of the Pollyannaish sort that seemed forced or inappropriate. Isabel simply refused to permit any obstacles to deflect her from her goal. She believed that she could achieve great things through determination, discipline, perseverance, and especially persistent optimism, and she held the same expectation for those around her.

Isabel was born in 1884 on a 120-acre farm near the settlement of Bailey in north-central Iowa. She and her family moved a few years later to the nearby town of Riceville. In 1906, at a time when few girls went to college, she followed her brother Warren to Iowa State Normal School, soon to become Iowa State Teachers College and now the University of Northern Iowa, in Cedar Falls.

Isabel led a wonderfully enriching and exciting college life. She played center on the girls' basketball team, the Shakes. At

Isabel before her graduation from Iowa State Teachers College, ca. 1910. *Warren Flath collection*

the end of her junior year she was elected to a committee to organize senior class activities. She also participated in campus theatrical productions; in her junior year she portrayed Flora, Goddess of Flowers, in a production of M. Nataline Crumpton's 1890 play *Ceres*. The following year she played one of the four court ladies in *Hamlet*; they "were all beautifully gowned and made a graceful picture," according to the campus newspaper, the *Normal Eyte*. The review continued, "The year 1910 will long be remembered in our school for at least two events: The evolution into a college and the presentation by its students of the play of Hamlet." Isabel also pursued a number of musical activities, including concerts with Warren. One Wednesday evening in March 1910, for example, the two of them traveled to perform in the nearby town of La Porte, population twelve hundred, and returned to campus the next day. Isabel graduated that year with a bachelor's degree in music and German. She was twenty-two.

It must have seemed to this adventurous, talented young woman that the whole world of art, music, and culture was open to her—but she needed to earn a living. In that era, even educated women faced a highly restricted choice of jobs, mostly teaching or nursing, and Isabel chose teaching. She landed her first job teaching music and English at Guttenberg, Iowa, and the following year she taught at Spring Valley, Minnesota.

Isabel's life took an adventurous turn when her brother Mert, a railroad engineer who had settled in eastern Montana, wrote to her and suggested, "Come out and homestead. The government is giving land away." Isabel found the prospect irresistible, and she and her college friend Helen Craft took up adjacent claims fifteen miles from Rapelje, a remote community in the flatlands of Stillwater County, north of Yellowstone National Park. Homesteading rules required a claimant to live on the land for three years and make certain improvements, so Mert built a shack for them right on the boundary, half on Isabel's claim and half on Helen's, permitting them to live together and still fulfill the requirement. Isabel carried a .45 and killed seven rattlesnakes during her stay.

Both women needed to earn a living, so Isabel found a teaching job in the Rapelje school, while Helen landed a post in the opposite direction. Each one needed to return to the house every evening to maintain their required residence, and for the commute they got two horses from their nearest neighbor, the Molt Ranch, about seven miles away. Isabel was not an experienced horsewoman, but she was effusive in her praise for Preacher and Ginger. These horses were trained always to come home to the ranch, with or without a rider, so the

Isabel with Preacher and Ginger on her homestead near
Rapelje, Montana, ca. 1912. *Warren Flath collection*

rancher told the young women that if they ever got into trouble
they were to turn one of the horses loose; when it came in at
the ranch without a rider, the ranch hands would immediately
begin a search. Isabel and Helen also had one buggy, which
they took turns using on their trips to school. After three years
they proved up on their homesteads; Isabel owned hers for the
rest of her life.

She was now free to move on, and she sought a better teaching job. Stanley High School offered her a position, and in 1915
she arrived to teach music, geometry, girls' physical education,
and German, until it was banned during World War I. She also
coached the girls' basketball team. Irene ("Bird") Edwards, my
dad's younger sister, was the star center on the team, once scor-
ing twenty-two points in a 47–1 rout of Bowbells. For two years
Isabel's team went undefeated, and in their 1917–1918 season

Isabel with a pig she was raising, ca. 1912.
Warren Flath collection

they won the state championship in Minot, defeating Coopers-town 11–5. It was Stanley's only state title for eighty-one years until the 1999 Bluejays football team broke the string.

Isabel's life changed dramatically when she married G. O. Flath in December 1919 and moved in with the rest of the Flath clan in Anton's busy house. Scuttlebutt around town was that Isabel had really wanted M. G. rather than G. O., but she was determined to have a Flath. In her previous life, she had followed a kind of proto-feminist path, making her own decisions, playing sports, performing music and drama, working in a profession, and beginning to accumulate assets. As a married woman, however, she was expected to conform to the era's expectations for a more constrained life. The first sign of her change in status was the loss of her American citizenship. The Iowa farm girl, native-born to a family whose ancestors had fought in the American Revolution, was stripped of her citizenship because G. O. came from Canada; married women assumed the citizenship of their husbands. Oakley became a U.S. citizen in 1920, and Isabel regained her citizenship after a ten-month lapse.

More permanent was the loss of her career, for schoolteachers were obliged to quit their jobs when they married. Isabel found herself somewhat unexpectedly marooned in a small Dakota town without a profession, deprived of the music, drama, and other cultural pastimes that she enjoyed and excelled in. The young woman who had played in *Hamlet* was now stuck in a true hamlet, and the loss of opportunity and ambition must have seemed immense. One way or another, this was to be her future. How she viewed that future, what private doubts or demons she struggled with, she rarely showed.

In 1924, after four years of marriage, she traveled to Minneapolis to attend a concert on the world tour of internationally renowned Polish pianist, composer, and statesman Ignacy Jan Paderewski, a performance that thrilled her. It surely reminded her of her earlier life and her exhilaration in music and drama, and she must have realized that whatever its other rewards might be, Stanley would never offer an opportunity for the full expression of her talents. Perhaps this moment was the watershed, for rather than lamenting her situation, Isabel decided to find new outlets for her energy and to do what she could to engender the fullest expression of the talent in Stanley. To start, she organized a local MacDowell music club. For the next six decades Isabel would be a force of nature in Stanley. She became the community's major musical impresario, willing the town to have concerts and recitals, leading the church choir, and occasionally giving lessons, although her patience did have limits. She was adamant that all people, especially children, could make music. She was also active in the Daughters of the American Revolution and became the Grand

Isabel, second from right, with her class in Rapelje,
Montana, ca. 1912. *Warren Flath collection*

Worthy Matron of the state's Eastern Star, a Masonic auxil-
iary. With her erect bearing and cultured enunciation, "Grand
Worthy Matron" seemed a natural appellation; she was a for-
midable but friendly presence whom few in town wanted to
disappoint. And she had a profound impact on my family.

My own first encounter with Isabel—Mrs. Flath, to me—
was in summer Bible camp when I was about ten. The church
organized a half-day school for young children for a week each
summer as a supplement to the hour-long Sunday school that
we attended during the year. Isabel taught one part of it, which
involved a competitive quiz. She would call out a question—
Who was the Israelite king before David? What are the names
of the four gospels?—and the twelve or fifteen of us kids sitting
around the table would raise our hands if we knew the answer.
I was able to answer nearly every one, putting me in the point
lead, because I had just finished reading *Hurlbut's Story of the
Bible*, which included short dramatizations and pictures of all

the important Bible episodes. Answering was so easy that it almost felt like cheating. When Isabel asked, "Who was the innocent man upon whom God visited many terrible troubles?" My hand shot up and I answered immediately, "Job," pronouncing it to rhyme with "Bob." Mrs. Flath said nothing but looked a bit dismayed, and I could see I had disappointed her in some way. She corrected my pronunciation and awarded me the point, but such was her presence that I was distraught at failing her and I didn't volunteer another answer all day.

To my sister Clarice, Isabel was "a shining ray in our town, because she was just different from the rest of us. She came with a bit of culture and a bit of something else the rest of us didn't have. There were so many things that were eye-opening around her. How to have a lovely dinner. How to have a lovely concert. She was just filled with life, a real model for us." Clarice first encountered Isabel when she was four and Isabel taught Sunday School. The half-dozen or so children sat on little red chairs in an airless room in the church basement, and Mrs. Flath narrated stories from the Bible in full dramatic form. Her innate urge to excel, even for this small audience, made the tales of Samuel or Jonah come alive for the kids. She led them in songs, the children belting out the words to please her. Eighty years later, Clarice could still recite:

> Dare to be a Daniel,
> Dare to stand alone.
> Dare to have a purpose firm,
> Dare to make it known!

The song's lesson, imprinted so early, was one Clarice would carry with her throughout her life, notably in opposing the segregated and casually racist atmosphere of northern Virginia in the 1950s and 1960s and later befriending Central American immigrants and helping them learn English.

G. O. and Isabel lived on Main Street in one of the nicest houses in Stanley. Our families moved in different circles; we never invited each other over for dinner, for example. Nor

Isabel's friend Helen and the homestead shack that they
shared fifteen miles outside Rapelje, Montana, ca. 1912.
Warren Flath collection

did Isabel participate in morning coffee, a custom of the local
women, who visited each other's houses mid-morning. Given
the town's strong Norwegian influence, "coffee" generally also
meant fresh-baked cinnamon rolls, homemade doughnuts, or
other baked goods, as well. These women gathered in shifting
groups of two or three to socialize and exchange news before
getting on with the rest of their day, but Isabel refused to participate in this custom, presumably because she viewed such
activities as a superficial waste of time. Some saw her attitude
as haughty, and it put them off, but to most, it just became part
of her mystique.

My mother Winnie occasionally attended a club luncheon
at Isabel's house, but mainly she found Isabel's self-assurance

and decisiveness intimidating. Isabel had instructed everyone in town, "Do not call between 1:00 and 2:00 P.M. That's my time to rest." She would go upstairs, take a nap, and *take no calls*. Such behavior was without precedent in Stanley; when the phone rang, we thought it impolite not to answer it. The grandeur of taking no calls amazed us. Isabel was an excellent cook, but domestic skills were not important to her, another attitude that distinguished her from other Stanley women. She was also frugal. Roy described how Isabel had once arrived at the post office to mail a letter with a three-cent stamp affixed to it. Informed that she needed only a two-cent stamp, she had said, "Oh, I'll take it back then." She promptly walked home, steamed off the stamp, and returned with the correct postage. Roy shook his head, bemused: "They've got a lot of money, why didn't she just send the letter and say, 'Next time I'll put a two-cent stamp on it'?" But Isabel did not believe in waste of any kind. And whatever the truth to the gossip that she had initially wanted M. G., she obviously loved G. O. She once told a teenaged Clarice, "Never to go bed without telling your husband you love him."

For whatever reason, Isabel approved of the Edwards family and decided that Clarice was a suitable playmate for her only child, a son named Warren whom she and G. O. adopted when they were in their late thirties. As a child Clarice was often invited to the Flath house. She arrived at two or three in the afternoon, and she and Warren played, sometimes taking a game from a sunroom closet that was stuffed with all the latest board games. Once Isabel planned an outing for the four of them to see a movie in Minot, a musical with a shallow plot contrived only to feature the singing of Nelson Eddy and Jeanette MacDonald. It hardly ranked with the great Paderewski concert, but Eddy and MacDonald were a step up from the restricted cultural offerings of Stanley. Clarice marveled at the elegance of the trip, from dressing up to riding in the Flath's luxurious car to the experience of the theater itself.

My sister Evelyn was in the Rainbow Girls, the youth affil-

Stanley's 1917–1918 championship girls' basketball team, coached by Isabel, top right. Roy's younger sister Irene ("Bird") Edwards is second from the right. *Warren Flath collection*

iate of Isabel's Eastern Star organization. Like the adult organization, the Rainbow Girls was a "secret" organization, and the girls were sworn not to reveal its ceremonies and rituals. They met upstairs in the Memorial Building, a community center, with a guard stationed at the door to prevent anyone from wandering in and seeing the proceedings. Existing members elected new members by putting a white or black marble in a box; even one black marble meant that the candidate was denied admission. Elections were also used to elevate members from one rank, or "station of the cross," to another. It was mainly a service club, organizing the girls to take on projects to help the community's poor or sick. Every Christmas they got big boxes of holly from Washington state, which the girls used to make wreaths. They then sold the decorations door-to-door, using the proceeds to buy food for poor families during the holidays. These projects were both worthy and fun; only later did an organization that literally used the black ball as its admission device strike Evelyn as unseemly.

In high school, Evelyn became Warren's girlfriend, and she would sometimes be invited to dinner at the Flaths'. Isabel served an elegant meal, with nice table settings, good food, and lighted candles. After dinner they would play bridge. All the Flaths, including Warren, were excellent bridge players. Evelyn was not; in our house, we played canasta, a distinctly more plebian game. Although Evelyn knew the rules of bridge, she was at sea when one of the Flaths, having bid six hearts, would lay down the cards after the first trick and say, "Okay, I have the rest of them." Evelyn thought, "How could that be, when I'm holding an ace?" They were right, of course, but they never criticized her play, even when she made a bad bid or a foolish lead. Later, after she became an excellent player herself, Evelyn realized that those evenings must have been tedious for the Flaths.

Evelyn's relationship with Warren continued through the first year or two of college, when she was at Macalester and he at Jamestown. One year when Evelyn came back from Saint Paul, the Flaths were planning a trip to Canada, and Warren insisted that Evelyn come along. They were gone for a couple of weeks, and although she was a bit uncomfortable with the arrangement, Evelyn received royal treatment not only from G. O. and Isabel but from all the Flath relatives they visited along the way.

Isabel, like the other Flaths in their different ways, lived a life of self-sacrificing service to the people who lived around her; without a whiff of self-promotion, her unflagging and dauntless optimism inspired others. She dared to be a Daniel. We noticed this trait directly, because her attitude seemed to say, "Those Edwards children look to be alright; I'll shape them into something." Patronizing as it sounds, we were grateful for her patronage. Our parents, indifferently educated themselves, revered learning and saw Isabel's efforts as opportunities for us and for the whole community. "Isabel was a person who changes your life," Clarice remembered, "always wanting you to reach for bigger things, believing there was another life outside the everyday life of Stanley. In fact many years later I wrote her a letter expressing my admiration and gratitude to her and remembering that little song we sang, because it's a good enough philosophy of life in four lines [and] you don't need anything else. But I never sent it, which I regret."

A SPIRIT OF ADVENTURE

. . . .

THE MIGHTY MO

. . . .

One popular image of folks on the Great Plains is of a staid, possibly joyless, and definitely unadventurous people. Leaving aside any debate over the aptness of this characterization today, it certainly doesn't fit our ancestors. The very act of homesteading in this region required a craving for adventure, because settling on the dry, desolate plains meant taking an enormous risk. In 1902, when the Edwards family disembarked from a "settlers' train" in Stanley with all their meager goods, farm machinery, and livestock, they knew they were intentionally leaving easier surroundings and putting themselves in difficult straits for years to come. Many settlers saw their lives as an adventure, and perhaps this attitude helped them to endure the tedium and grind. Elizabeth Corey was a single Iowa farm girl who successfully homesteaded west of Fort Pierre, South Dakota; her letters home reveal her indomitable spirit of adventure despite the terrible hardships and setbacks she endured. Elinore Pruitt Stewart, a single woman who homesteaded in Wyoming and married her neighbor, fired the public imagination with her sense of adventure in reports to the *Atlantic Monthly*. Isabel Flath displayed the same high spirits and desire for adventure when she homesteaded in Montana. When these settlers or their descendants moved to town, they kept their taste for excitement, as when Roy built his house trailer or when Clarice went off to the shipyard. Even though most had limited means and possessions, they

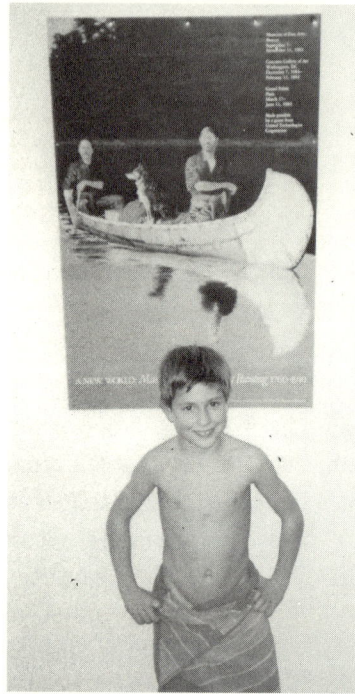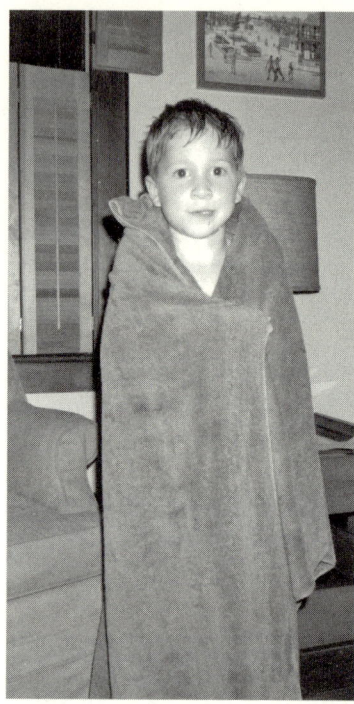

Sam, the senior member of the crew at eleven years of age,
in front of the poster that inspired our canoe trip; and
George, the junior mate at eight years of age, Spring 1990.
Richard Edwards collection

delighted in finding ways to create amusement and adventure
for themselves.

In the spring of 1990, feeling this urge for adventure, I
hatched a plan with my two sons, eleven-year-old Sam and
eight-year-old George, to canoe the Missouri River. We lived
in Massachusetts with my wife and our baby daughter, and
it was our habit for the boys and me to go out west for some
skiing or dude-ranching each year. This year was different, because I planned to return to North Dakota to introduce the
boys to the wide-open prairie of their ancestors. We wouldn't
get to Stanley, but we would be nearby, on the Missouri. Our
inspiration came from a Frederic Remington poster we passed

each day in our back stairway showing two intrepid *voyageurs* and their dog, alert and watchful on a lake in Canada. We too would be explorers, and we could test how the Old Stanley spirit of adventure had been passed down (or more likely watered down) through the succeeding generations.

I'd done a fair amount of springtime canoeing in New England with my friend Sam Bowles. Sam was much more skilled than I, and he always took the stern. One February we canoed the Westfield River in western Massachusetts, which is shallow and smooth in the summer but in springtime is a raging torrent—a whitewater canoeist's delight, but a deathtrap where it plunges through the deep Chesterfield Gorge. The river was surging that day, and as we came closer to the gorge, we passed a couple of possible pull-out spots and thought that we probably ought to be stopping someplace soon, but we were enjoying the small rapids too much to quit just yet. Suddenly, however, the current got much stronger, and we began to have trouble handling it. We couldn't see the gorge yet, but its tremendous roar said that it was very near. At the last possible moment, Sam lunged for some bushes hanging out over the water and held on tight, saving us from being swept right through the falls. We were shaken but exhilarated, too.

I wanted to share a less dangerous version of this adventure with my sons, and I also wanted them to connect with North Dakota. I remembered my boyhood in Stanley and vaguely recalled tales of the construction of Garrison Dam and moving a town that was to be flooded. I got out a map and saw a long, enticing stretch of the Missouri running from below the dam all the way south to Bismarck. We would be canoeing in the summer, long after any spring floods were over, and the flat terrain of the plains would produce a placid, slow-moving river nothing like the swift and dangerous streams of New England. It seemed ideal for two inexperienced young boys and their dad. What an adventure: we would be floating down the Mighty Mo! The Big Muddy! Retracing the path of Lewis and Clark! (Never mind that we would be going downstream, while they had to struggle against the current. Two of us were children, after all.)

Our first evening on the river seemed to confirm what a good idea this adventure had been. There was a beautiful sunset over the prairie. We stopped before twilight and set up our tent on a little bluff; the land was mostly flat, and we could see for miles. When we tucked into our new sleeping bags, we were happy and almost too excited to sleep. Eventually, however, the day's exertions and the evening's deepening darkness worked their will, and we dropped off.

The sound of heavy rain awakened me, and soon the sky was alive with lightning. The thunder deafened us with loud, repeated cracks. The boys had woken up before me, and they thought the storm was grand fun. They watched the lightning flashes light up our little lime-green nylon bubble of a tent, making it glow like the mantle in a camp lantern. Then the bubble burst, the tent blew down, and the rain and sand came in.

And that was the first time I wondered: *What kind of father brings two innocent children into this mess?*

Before the Internet, it was difficult to research the conditions on a river. I had sent away for information from North Dakota's state tourism department, and I received a packet with photographs showing smiling boaters and water skiers on beautiful sunny summer days enjoying the unexpected treasures of the northern plains. Oddly, there were no pictures of violent summer storms. I wrote to Jack's River Trips, one of the canoeing outfitters listed in the brochure, and Jack agreed to supply us with a canoe, paddles, and life jackets, and to drop us off where we'd put in the river and pick us up where we came out. We seemed all set.

Neither the tourism department nor Jack gave a disclaimer, but this description of the river from Pierre-Jean De Smet, a famed nineteenth-century missionary and traveler, would have been quite useful: "I fear the sea, I will admit," the world-wise Jesuit wrote, "but all the storms and other unpleasant things I have experienced in four different voyages did not inspire so much terror in me as the navigation of the somber, treacherous, and muddy Missouri."

I could have made good use of Father De Smet's opinion, but instead we spent the spring buying new sleeping bags, planning meals, and equipping ourselves with knives, fishing poles, dishes—all the things we'd need. We had a well-used JanSport tent with sturdy fiberglass poles and a fly, just the right size for the three of us. Each of us had a backpack packed with his own clothes and some of our shared gear and food. We planned to be on the river for three days and three nights. We'd put in just below Garrison Dam. Using a ruler on the North Dakota road map, I figured the river distance from the dam to Bismarck to be about sixty miles. Floating with the current, twenty miles per day seemed reasonable.

July arrived and we flew to Bismarck, where we stopped to visit Irene and Arne before beginning our adventure the next day. Here we also picked up the trail of Lewis and Clark and their Corps of Discovery. They had begun their journey at Saint Charles, Missouri, in May 1804, and by late October they had paddled, poled, and pulled their boats nearly a thousand miles up the Missouri to a place near present-day Bismarck. There they found extensive Indian villages where more than four thousand Mandans, Hidatsas, and Arikaras lived, including approximately one thousand warriors. These were sedentary, agricultural peoples who participated in a vast trade network that extended all the way to the Pacific coast. The forty-man corps built a stockade, Fort Mandan, and that winter they traded and hunted with the Indians and obtained critical information about the regions and peoples they would encounter on their travel west. It was here, too, that captains Lewis and Clark engaged their two interpreters, a Frenchman named Toussaint Charbonneau and his teenaged Lemhi Shoshone wife, Sacagawea, also called Sakakawea.

Sam, George, and I spent a late afternoon and evening with Irene and Arne, describing our plans and proudly showing off our new equipment. I hadn't seen Irene for years, I barely knew Arne at all, and they had never seen the boys, so everyone enjoyed getting to know each other. They were living in their splendid new house on the Missouri. Their yard ran from the house down to the river, with a big garden at the water's edge. Anything planted in that rich black soil grows wondrously if it can find moisture, and irrigating from the river, they grew beautiful rhubarb and vegetables of all kinds. George and Sam ran around the yard and played with four six-week-old kittens.

After dinner, Arne and I strolled down to the river, where we stood for a few minutes in the gathering dusk. A soft-spoken man, never one to stick his nose into other people's business, he nevertheless said, "Rick, be careful out there with the boys. This river is deceptive and not to be trusted. The sand bars and the currents are always shifting, and it's tricky and dangerous."

"We sure will be careful," I assured him, but of course I was thinking, "Oh, we'll be *fine* on this river. I've got experience, I know what I'm doing."

The Corps of Discovery left Fort Mandan on April 7, 1805, and headed upriver in two pirogues and six canoes. That day Meriwether Lewis wrote, "I could but esteem this moment of my departure as among the most happy of my life. The party are in excellent health and sperits, zealously attatched to the enterprise, and anxious to proceed; not a whisper of murmur or discontent to be heard among them, but all act in unison, and with the most perfect harmony." I felt the same about my own party of adventurers.

Lewis and Clark made excellent progress on the river, twenty-five miles or more on good days, and soon they approached grizzly country. The Indians had warned them about grizzly bears, but Lewis was unconcerned. While grizzlies might pose a real danger to men armed only with bows, he thought, they would prove no match for a man with a rifle. One day Lewis and several of his men spotted a pair of grizzlies. They wounded one of the bears, which managed to escape, but the other bear ran directly at Lewis, chasing him some eighty yards; he and one of his men escaped only when they managed to reload and kill the bear. Lewis should have had more respect for the dangers of a country that he knew little about—but then, he wouldn't be the only information-challenged traveler on that river.

The next day we went to meet our outfitter, Jack, and make the final arrangements. Sam and George were especially proud of their new fishing licenses—fishing had been a frequent topic of discussion that spring. Eager to get going, we decided to have Jack take us out to the river that very afternoon, giving us an extra night to camp out. Three hours later, around six o'clock, we arrived at our put-in spot a mile or two below Garrison Dam. Later we learned that Lewis and Clark had camped in this area on April 8 and 9, 1805. It was also precisely the area where, in Clark's words, "one Canoe filed with water every thing in her got wet. ⅔ of a barrel of powder lost by this accedent." Unaware of all this history, we put our canoe in the water, loaded the baggage, cinched up our life jackets, crawled aboard, and pushed off. Waving goodbye to Jack, I told him we'd see him in Bismarck. It was just as I had dreamed: smooth water, a glorious sunset behind us, and an open river in front of us, just me and my boys, completely alone on the river. There was not a man-made thing in sight. What a beautiful trip this was going to be!

We canoed for maybe an hour and a half. Then the boys got hungry, or at least eager to stop and try out all our new cooking, camping, and fishing gear. We pulled over at a welcoming spot where there were sandbars in the middle of the stream and a small sandy beach at the riverbank. The only unpleasant surprise was the temperature of the water. The dam impounds chilly snowmelt in the spring, and because the power plant's intake structures are at the bottom of the dam, it discharges the coldest water from the reservoir into the river. During summer, the water temperature on the river can be as low as fifty-five degrees, making it way too cold to stay in for more

Meriwether Lewis stands on the bank of the Missouri River while his experienced rivermen struggle as "one Canoe filed with water" in 1805. The Edwards rivermen struggled to avoid being tipped over by the wind on the same stretch of river. *Art by Dustin Young; Richard Edwards collection*

than a few minutes. Moreover, the current seemed to be running pretty fast. Taken together, these discoveries meant that we wouldn't be doing any swimming by choice, and since the boys were not strong swimmers, the prospect of tipping the canoe over began to appear more dangerous.

We got to work setting up camp. After debating whether to camp on one of the sandbar islands in the middle of the river, we decided that we liked the little beach along the shore better. It had twenty or thirty feet of sand and gravel, and behind that a dirt shelf rose up, perhaps three feet high. I fired up our small cookstove to boil water for the inevitable macaroni and cheese, and the boys started to set up the tent, just as we had practiced at home. The obvious place to pitch it was on the little shelf, but then I noticed a small bluff set back from the beach, about fifteen feet above us. I thought it might be neat to be up there where we would have a better view. The boys liked the plan, too, and soon they had the tent all set for the night.

George was eager to try his new fishing rod. His grandfather Roy had been a great fisherman who caught fish all over the upper plains and in the Montana mountains, but they say that the good-fishing gene sometimes skips a generation, and indeed I am a poor and impatient angler. George might have Roy's fishing gene, but he had a poor mentor, and in any event I was too busy getting dinner to supervise. There he stood on the bank, a bundle of eight-year-old energy who couldn't wait to get his hook in the water and catch a big one. He pulled his rod back, just as we had practiced in our backyard, and flung it forward to make a big cast out into the river. Unfortunately, he forgot to hang on to the rod. The whole thing—hook, line, rod, and reel—flew off into the cold, fast-moving, murky water and was gone. We hadn't been in camp more than fifteen minutes, and George's fishing experience was over. Perhaps it was an omen of things to come.

After dinner and cleanup, we were set for the night. Well, almost set. Somewhat to the boys' annoyance, I insisted that we pack up our gear and put it back in the canoe in case any critters came by; pick the life jackets up off the sand where we dropped them and buckle them around the thwarts of the canoe; and then pull the canoe ten or twelve feet up on the sand. Out of an excess of caution even I knew was silly, I tied the painter line of the now well-beached canoe to a downed tree trunk lying well back from the shore. And despite the clear sky, we even put the fly up over our tent, more for practice than anything else—on the first night, using our gear was fun, and a tidy camp is just a good habit. Finally we turned in for the night. Aside from discovering the chill of the water and the strength of the current, and the loss of George's fishing rod, it had been an excellent first day. Well, an excellent two hours.

This is the point, of course, where you came in.

We woke up in the middle of the storm. The rain was heavy as only prairie downpours can be, but that wouldn't be a problem. Our tent was designed to take it. It was not designed to withstand lightning, which came with sharp, deafening cracks, again and again and again, lighting up our tent like a strobe. The boys watched the top of the tent with glee, calling it "old veiny," because the patterns that the lightning drew there looked like details of the circulatory system. I worried that our tent, sitting high up on the bluff in a spot chosen for its superb view, was now ideally placed to become a lightning rod. Then the wind came up, roaring and driving the rain almost sideways. The tent fly wasn't so effective against horizontal rain, but that concern turned out to be minor, for soon the wind blew it off the tent altogether. Now the rain and lightning felt much closer, and water began to accumulate inside the tent. Finally the tent itself blew down on us with a staccato crack that could only mean that the fiberglass poles had broken. The rain poured in, and a thought flickered across my mind: "What kind of stupid idea was this, bringing two small boys out here in the middle of this hell?"

We spent the rest of the night in the mostly-collapsed tent, which lay over us like a leaky tarpaulin. We managed to keep most of the rain out. Sam discovered that it was better to sleep *under* your sleeping bag, using it as a shield. We awoke to a chilly, overcast, drizzly, unpleasant kind of day. But the most astonishing sight was our canoe. It was floating in several feet of water. Where was the sandy beach? Gone. Where were the sandbar islands in the middle of the river? Gone. Instead the river stretched uninterruptedly from bluff to bluff. The canoe was floating placidly, tethered by the painter to the log.

When I saw the scene before us, my mind exploded with a

litany of horrible what-ifs. What if we hadn't put all our gear, including our life jackets and paddles, back into the canoe? Everything would have been washed down the river, and that would have been that. What if at the last minute I hadn't tied our well-beached canoe to the log? Then the canoe, too, would have been lost. *What if we had camped on the sandbars in the middle of the river?* How could we have escaped in the dark of night, in the rain and the wind and the lightning, with the swift, cold river swirling and rising all around us? This string of miraculous near-misses had to mean that someone somewhere was looking out for us, because I was sure as hell out of *my* league.

I got up and ran over to the downed tree, grabbed the painter, and pulled our canoe back up on the new shore. Okay. Get a grip, Rick. We're wet, our tent is broken, it's chilly out here, but we're alive, we have our canoe and all our gear—well, except for George's fishing rod—and we're okay. Just put the nightmare image out of your mind. Forget it.

If I could draw on some dauntless optimism, things might not seem so bad. We found the tent fly some fifty yards away. Only two of the fiberglass tent poles had broken, and fortunately I had brought two extras along. All our gear was safe, and although our life jackets were soaking wet and would be uncomfortable to wear, our other stuff, including our food, had been stored in plastic garbage bags inside our backpacks and was not even wet.

I was surprised that even an intense storm like that one could elevate the river so quickly. The Missouri this morning was six or eight feet higher than it had been last night, and in my mind I bitterly resented that no one had warned me how tricky and dangerous the river was. Well, except Arne. Maybe what I really needed was more Old Stanley modesty about my river expertise. I later learned, however, that the storm alone was not responsible for the change in water level. It turned out that the Army Corps of Engineers often released water from

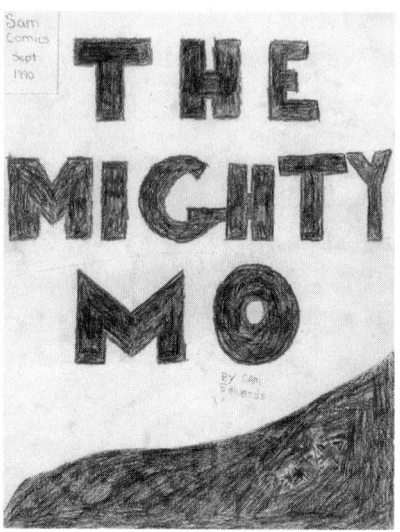

Two pages from *The Mighty Mo*, Sam's comic book retelling our Missouri River adventure. *Richard Edwards collection*

NATIVES OF A DRY PLACE

the dam in pulses, and where we were, just a few miles below the dam, those releases regularly raised the water level eight feet or more, temporarily submerging all the sandbars.

You'd think I would have known more about Garrison Dam. After all, it had been built between 1947 and 1953, when my family and I were living in Stanley, just fifty miles to the north. Its construction was part of a dam-building frenzy that lasted from the 1930s through the 1970s, when virtually any stream in the Great Plains or mountain West was dammed, channelized, or otherwise tamed. Eventually the public came to understand that free-flowing rivers and all their attendant inconveniences, including spring floods and periods of low flows, were critical to the region's biodiversity, and resistance to new dams grew. By then, however, enormous dams already dotted the western landscape. On the Missouri River alone there were six major dams: Fort Peck, Garrison, Oahe, Fort Randall, Big Bend, and Gavins Point. Indeed, we were paddling on one of only two remnants of the free-flowing river left in North Dakota. In other places along the Missouri, the Corps constructed hundreds of miles of levees to channelize the river, effectively turning it into a drainage ditch.

Garrison Dam was initially promoted as a panacea for the region's ills, with benefits for everyone. Folks in Omaha and Saint Louis would benefit from barge navigation and flood control. Flooding was fresh on everyone's mind because the 1943 Omaha flood had been particularly devastating. Further downriver, the rampaging Missouri had leaped cross-country to join the Mississippi six miles below Saint Charles, Missouri, instead of at its usual meeting place twenty-two miles further south, and one thousand people were stranded in the triangle thus created. People in the Dakotas and further upstream were told they'd benefit from cheap hydroelectric power, irrigation for vast fields, recreational opportunities, and urban water supplies. Besides, most people thought, the land to be flooded was mostly Indian land, and everyone accepted the idea that they didn't do much productive with it.

Garrison was a huge project. It required nine million truck-loads of dirt and 1.5 million cubic yards of concrete, more than twenty times more material than was used to build the Great Pyramid of Giza. The dam's embankment is two and a half miles long and 2,050 feet wide at its base. Its construction encouraged the development of the enormous trucks and dragline excavators now common in open-pit coal and copper mines. The lake behind Garrison Dam covers nearly four hundred thousand acres of rich, deep, black alluvial soil, the product of millennia of prairie growth and decay, the kind of soil you pay ten dollars a bag for at Home Depot. This river-bottom land was once one of the most diverse and productive ecosystems in the whole Missouri watershed, a rich riverine forest of cottonwoods, willows, and dogwoods.

My family and I did not pay much attention to Garrison Dam while it was being built or in the years afterwards as the land behind it slowly disappeared. I remember hearing about how they had to move the town of Sanish; they dragged many of its wood-frame houses and businesses up on a bluff where they'd be out of the reach of the advancing lake and made a new town, which they inventively named New Town. Sanish had been the site of the only bridge over the Missouri for miles around. Irene and Arne used to go picnicking down there and took glamorous, self-dramatizing photographs of themselves on that bridge. Two other towns, Van Hook and Elbowoods, were also flooded under.

Only later, when American Indians gained a stronger voice, did the destructive legacies of this mammoth re-engineering of the river come into clear focus. One was quantitative. The Fort Laramie Treaty of 1851 had granted the Arikara (also known as Sahnish), Mandan, and Hidatsa tribes a reservation of more than twelve million acres. Various land-grabs and redefinitions of the law had reduced the reservation to three million acres by 1880, and then a presidential executive order in 1910 cut it back further to 640,000 acres. The lake behind Garrison Dam deprived the Indians of another 152,000 acres of their best land. But the loss was qualitative, as well. The river valley had been home to the Three Affiliated Tribes for hundreds of years. They had adopted ways of herding, farming, and hunting that suited this particularly fertile environment and developed emotional and religious ties to the land. The Garrison Dam destroyed their communities and way of life and dispersed their people. As the powerful film *Waterbuster* shows, many Indians moved to other cities and states, some as far away as California, losing any connection to this ancestral place and their creation myth. Some Indians bitterly suggested that the new town being built up on the bluff should commemorate the loss of both Van Hook and Sanish and be called Vanish.

Two ironies cap the story of Garrison Dam. First, although flood control was the most urgent reason advanced for its construction, a smaller dam would have answered this purpose. The massive scale of the project was dictated by a requirement to support navigation far downstream, but competition from trucks and railroads has made barge traffic essentially irrelevant. Second, the Corps of Engineers chose to name the new reservoir Lake Sakakawea, to honor the region's Indian heritage.

Sam, George, and I were ignorant of this bitter history. We did know that we'd survived the big storm that had blown down our tent, and we knew that we had to keep moving down the river. We cooked our morning oatmeal and hot chocolate. George and Sam were still in good spirits, but I was a bit depressed and anxious about the trek ahead of us. We packed up our wet tent and other gear, loaded it all into the canoe, and got ready to set off again. This morning's canoeing seemed a lot more serious task than last night's pleasurable float.

A gusty and somewhat drizzly wind blew directly towards us from across the river, whipping up little whitecaps on the water. We stepped into the canoe, Sam in the bow, George in the middle, and me in the stern. I pushed us off from the bank. We made it perhaps ten or fifteen feet out into the river, and then the wind caught the canoe and started to push us sideways. "Keep us pointed right into the wind," I yelled to Sam. "We need to get farther out in the river." He tried but couldn't hold the bow into the wind. We got turned sideways, and then the wind really caught us. I worried that we might tip over, so I steered us toward shore. Back on the bank and safe for the moment, we discussed our strategy.

Both boys could see now that I was really worried. I'd have worried even more had I known we were near the spot where Lewis and Clark had "one Canoe filed with water." If their experienced rivermen had swamped a canoe here, what chance did we have? The boys were resolutely determined to help, and we tried again. This time we got twenty or twenty-five feet out into the river, but then the wind caught us again. Sam struggled but couldn't hold the bow straight. Again we turned sideways and felt the full force of the wind pushing against the canoe. If we swamped in this cold water with the strong current, how would I even get the boys back on shore?

We went back to the bank and decided to wait for the wind to die down. We waited perhaps an hour, maybe longer. In our enforced idleness I entertained a new worry: we needed to make at least twenty miles on the river every day, and we were making none just sitting there. Although the wind had not quieted much, I got impatient and decided to start again. This time I put myself in the bow and Sam in the stern. He pushed us, and we headed out into the river. I was able to keep the bow into the wind, and we began to make progress. We still were in danger of being pushed sideways and even swamping, but we learned how to control the canoe in the wind.

When we had journeyed a mile or two, the river turned a bend, and now the wind came at us over our left shoulders, still pushing us sideways but more diagonally. Presently a power plant or some kind of industrial facility came into view on the right bank, exactly in the direction the wind was trying to push us. A big open water intake stared ominously at us, covered only by a grate. Worried about this new hazard, we steered further out into the river to give it a wide berth. The river here was wide and slate gray, and it seemed a long way to shore and safety. Gradually, however, the wind abated, blue patches appeared in the sky, and the sun came peeping out. The air warmed up, and the river seemed less turbulent. We passed the evil-looking intake, and life looked up.

By the time we stopped for dinner, the weather had regained its warm, sunny disposition of the previous evening. We found a *safe* campsite—well up and away from the river—and began unloading our wet gear. We got our tent up again and mounted the fly to dry it out, then began thinking about dinner. We had mostly brought dried meals, but we also had a canned ham—in our view, the prize of our field cuisine. Given our miserable night and hard day, I announced that we deserved the ham now, and the boys reacted with their customary enthusiasm. George went to his backpack and proudly produced the ham, which had been one of his special responsibilities.

It was fairly heavy, sealed in one of those cans shaped vaguely like a heart. To open it, you used the little key glued to the underside to unpeel the metal ribbon that ran around the top. I turned the can over to retrieve the key, but it was gone. I asked George about it, but he had no idea that such a key even existed. I figured it had probably just fallen off in his backpack, so we looked there. No key. We searched again, more carefully. Still no key. George cried a bit, thinking I was angry with him, but I assured him the loss wasn't his fault. Back in our living room, that ham and our other gear had been endlessly packed and repacked, debated over, inspected, and tried out, as the boys impatiently waited for the day when the trip would begin. That ham had undoubtedly been in and out of all three backpacks, so we searched them all, but still no key. Finally I plied the can with my Swiss Army knife for a good fifteen minutes, but without the key, the damn thing might as well have been a Diebold safe. We never did get it open. George carried it with us all the way back to Amherst.

The next couple of days were uneventful and pleasant. The

weather was warm and sunny, even hot during the afternoons. The river was placid. On our last day, the boys trailed their feet in the water, enjoying its coolness contrasted with the heat of the day. We stopped on sandbars a few times to play. It was marvelous, but there was one nagging concern: we could not really measure our progress down the river. In three days, we had seen no people or boats. There were almost no landmarks. No towns, bridges, or mountains off in the distance to tell us where we were.

About halfway between the dam and Bismarck, the map showed a highway bridge that crossed the river at the little town of Washburn. Had we been on schedule, we should have hit it about midday on our second full day, or even earlier if we factored in the paddling we'd done on the first evening. But when we camped after two full days of paddling, we still hadn't seen a bridge. Had I misread the map, and there *was* no bridge? Where could it be? We continued downriver the next day, having a wonderful time, but I grew increasingly concerned that I had drastically miscalculated the distance. Early in the afternoon of our third day, we turned a bend, and there was the bridge. We'd been on the water almost the three full days we'd scheduled and apparently had made it only halfway.

This revelation was discouraging. We had enough food for four days, maybe five, but at this pace it would take us six days to reach Bismarck. I had planned a trip to Theodore Roosevelt National Park for the next three days, and Arne had arranged a hunting cabin for us to stay in. Obviously we couldn't do that if we were still on the river. Besides, where would Jack the outfitter be, and where would he meet us? We could see a few lonely buildings up near the bridge, presumably the town of Washburn, but the map showed no other towns along the river all the way to Bismarck, so we had to decide then.

We held a brief consultation and collectively decided that we should stop here at Washburn and see if we could get Jack to come and meet us. We pulled the canoe up under the bridge and climbed onto the roadway. Washburn turned out to be a tiny crossroads, with only a few businesses and a few houses. It was Sunday afternoon, and everything was closed; neither cars nor people stirred. Washburn appeared to be asleep. I walked down the two blocks of deserted main street and spotted a phone booth. Fortunately I had some change, and I dialed Jack's number. He answered right away. "Yeah, I was wondering how you guys were doing out on the river. You okay? That was quite a storm we had, wasn't it! The news people here [in Bismarck] reported on all the downed power lines and damage to roofs. Must have been something out there on the prairie."

I explained that we were in Washburn and were ready to be picked up. Jack said he'd be out to get us right away but that it would take him a couple of hours to reach us. The waiting was painful, knowing that our adventure on the Missouri had come to an end and that we'd failed to reach our goal. Finally he pulled up, and we chucked our gear into his pickup and rode back to Bismarck. We slept at Irene and Arne's house that night, and they were the first to hear our tale of the big storm on the Mighty Mo, a tale that has grown more dramatic with each retelling.

So how did we do, measuring our spirit of adventure—bolstered, I should say, by resoluteness and pluck—against that of our ancestors? We survived a couple of crises on the open landscape and generally held cheery attitudes while doing it, keeping our canoe upright and ourselves out of the treacherously cold water. On the other hand, our experience in a sometimes hostile environment had lasted four days, not the thirty years or more that most early settlers faced before achieving more comfortable lives for themselves. And unlike them, we could call Jack to rescue us when we'd had enough.

Well. I'd say perhaps we did alright by Massachusetts standards.

MODESTY

· · · · · · · · · ·
A CODA
· · · · · · · · · ·

Old Stanley values and habits of mind persist, and when they are forgotten, sometimes they must be relearned. In 1981, my family and I summered at our house on Martha's Vineyard. Back in March, I had been invited to lead an all-day workshop for high-school history and social science teachers at Phillips Exeter Academy in Exeter, New Hampshire. The event was part of a week-long institute, and they planned to bring in a different facilitator each day. I was to be the featured bigwig on a Wednesday late in July.

I told Exeter that I was happy to come, especially since they offered a rather handsome honorarium, but I pointed out a logistical difficulty. If I had been at my house in Amherst, Massachusetts, driving the couple of hours north to Exeter would have been no problem. But in July I would be coming from the Vineyard, and with the extra driving and the usual long wait at the ferry, the trip would be much longer. Consequently, I accepted only on the condition that they buy me an airline ticket from the Vineyard to Boston's Logan Airport and have someone meet me there to drive me up to Exeter and back. I was pleased—and a little surprised—when they agreed.

By this time I thought I had moved far beyond my Stanley origins. After graduating from Grinnell College, I went to Harvard, receiving my doctorate in economics in 1974. I took up a teaching post at the University of Massachusetts in Amherst, where I became a tenured associate professor. I spent my first sabbatical building our house on Tisbury Great Pond on the Vineyard.

Modesty was a virtue both prized and expected in Old Stanley. Pretension, exaggerated self-promotion, boasting, or putting on airs was thought to be coarse, a kind of social lapse like farting in public. Self-praise is no praise at all, they said. If you were good, others would do the praising; if you weren't, praising yourself would convince no one. Of course, you rarely heard anyone challenge a person who had a puffed-up view of himself, because that, too, would have been awkward—but townsfolk knew who the pompous and conceited people were.

The expectation of modesty also extended to discussions about one's family. Although Roy and Winnie were pleased with their children, they would never directly say so to their friends, because that would seem like showing off. If someone said about a teenaged Evelyn, "Well, you've got a really smart girl there, she's a special one," Roy might respond, "Well, she's still got some things to learn in the kitchen, she's pretty slow on those dishes." To deflect the compliment in this way didn't mean that he wasn't a proud parent, and the emphasis on modesty never inhibited anyone's genuine feeling of self-worth. It did, however, provide a healthy check to smugness and superiority. It was obvious to everyone that Isabel Flath was among the best-educated and most talented people in town, but neither she nor her family would ever have said so publicly.

Eventually the day of my Exeter performance arrived. I got up early and drove over to the tiny Vineyard airport for the morning flight. These were the days before airport security got serious; you showed up, they checked your ticket, and when the plane was ready you walked out onto the tarmac and climbed aboard. They did ask how much you weighed, because the plane was so small that they needed to distribute the load properly. One rather stout passenger clearly did not want to divulge her weight, but the gate agent insisted, so she told him in a voice so quiet I had to strain to hear. Then, in a stage whisper loud enough for everyone to catch, she told her traveling companion that she had never before endured such an outrage.

When I presented my ticket, I noticed that the agent examined it a bit longer than the others. He rechecked the passenger roster pinned beneath his clipboard and looked through some other papers. I was about to ask him if there was a problem, but then he scratched a big "OK" on my ticket and handed it back, so I took a seat and waited. Soon our plane arrived, the inbound passengers descended, and we boarded. The forty-five-minute flight to Boston was uneventful.

At that time, Vineyard flights didn't unload at the big Logan terminal but instead discharged their passengers into a small, dingy building at ground level. I entered the little waiting room and looked for the Exeter driver, but there was no one waiting for me. Disappointed, I took a seat on one of the battered pastel-colored hard plastic chairs lined up against the wall. I waited ten or fifteen minutes, wondering what had happened to the driver. Stuck in traffic, I supposed. The waiting room cleared out, and soon the only people left were the two men behind the counter and me. The wait stretched to half an hour. At forty-five minutes, I began to get a little irritated. I wasn't John Kenneth Galbraith or Lester Thurow, but I wasn't chopped liver, either. I certainly deserved better treatment than I was receiving. After all, *they* were the ones who had asked *me* to be their featured scholar that day. The least they could do was to get their driver to the airport on time to pick me up.

When I had waited a full hour, I decided to take action. As it was before cell phones and email, I had to call Phillips Exeter using the pay phone hanging on the wall. After I was transferred to the correct department, a secretary answered. I identified myself and told her that I'd been waiting for their driver for more than an hour. "What's the problem?" I asked, keeping my tone civil but nonetheless conveying my irritation and the fact that I expected some quick remediation. She seemed a bit flummoxed, but after a short whispered conference with a colleague, she apologized for the confusion. She said that they would start their driver out for Logan now and that he would arrive in about an hour and a half. I kept my cool, but I also made it clear that I was not happy about the situation. Another hour and a half? I couldn't imagine what kind of excuse Exeter had for dropping the ball like this.

Really annoyed now, I sat down for the long wait. I hadn't expected to spend three hours hanging around this dingy little terminal with no coffee and nothing to eat but candy from the vending machine. I reviewed my notes just to relieve the boredom. I bought a newspaper. The waiting room began to fill up again with travelers taking the next flight to the Vineyard. Eventually they called the Vineyard passengers to board.

Idly, I looked at my plane ticket. Then I noticed something odd. The ticket was dated July 22. That seemed wrong. I checked the date on the newspaper and confirmed that it was, as I thought, the twenty-first. Then, with growing horror, I noticed that the paper also said *Tuesday*. *Tuesday*, July 21. But maybe the paper was a day old! I raced over to the counter and asked, "What's today's date?" The agent, busy checking in the last of the Vineyard-bound passengers, replied, "The twenty-first."

I had lost track of what day it was. That was easy to do during the idle summer days on the Vineyard, where except for Sunday and the Sunfish races, every day was more or less the same and my only appointment was an afternoon stroll to the beach. I had taken my flight a day early. No wonder the gate agent had puzzled over my ticket. No wonder there was no driver waiting for me. No wonder the secretary at Exeter was confused. And what did she think of my arch tone when I complained that there was no driver to meet me? What should I do now? Go back to the Vineyard and do this all again tomorrow?

I asked the agent what I could do. He told me that I could get on the flight just leaving, which was the last Vineyard flight of the day—but I would have to do it right away, because all the other passengers had already boarded. "Do I have time to make a quick phone call?" I asked. "No," he said, "the flight is ready to take off." If I didn't take it, I would be stuck on the mainland, so I got on board. On the way back to the Vineyard, I counted the losses. Beyond a day completely wasted, I had used the plane ticket that Exeter had purchased, so I'd need to buy another one tomorrow, cutting into that nice speaking fee that I had hoped to pocket. I had thoroughly annoyed the folks at Exeter. And now there was a driver on his way to Logan with no one there for him to pick up.

When I got off the plane at the Vineyard, I went straight to the pay phone and called Exeter. I got through to the secretary again and told her where I was. She was not amused, and received my news with barely-concealed irritation. I tried to explain that I had gotten confused about the day, that I was sorry, that I still planned to come the next day, and did I mention I was sorry? She told me that she had no way of contacting the driver. When he got to Logan and couldn't find me, he would probably call in, and she'd tell him to return to Exeter. Somewhat frostily, she said they'd look for me at Logan again the next day.

I did return the next day, and this time the travel arrangements went smoothly. When I arrived at Exeter, it was clear that all sixty or so of the assembled teachers knew about my previous day's misadventure. They were openly hostile, and it wasn't hard to imagine why. "We're giving up a week of our precious summer vacation for this Harvard guy from the Vineyard who can't keep his days straight? What could he possibly teach us?"

Fortunately for me, before they came to Exeter, these teachers had written short essays on their interests and what they hoped to gain from the institute. I had spent some time going through these papers, and I'd put together some common themes and matched the authors' names with them. I started the workshop by going around the room and having each teacher briefly describe what she or he wanted that day's work to achieve, and I made a comment or two for each teacher based on my reading of his or her essay. When they saw that I had not only read their essays but connected their names and themes—in short, that I'd done my homework—their hostility melted away. Apparently the prior visiting bigshots had not been so conscientious.

At Exeter, I was retaught a piece of Old Stanley wisdom, a kind of Third Law of Psychomechanics—or maybe it's just karma: for every pretentious, self-important, and overbearing action there is an equally nasty and opposite reaction. The person to whom you are curt or patronizing today is inevitably exactly the person from whom you'll need to ask a favor tomorrow. In a small town, this knowledge is a bit of survival wisdom, but for me at least, it seems to operate even in a larger, more anonymous society. At the very moment when you are most enjoying being condescending and supercilious, you might as well begin preparing your self-abasement. Don't forget the Old Stanley virtue of modesty, and you won't be embarrassed as I was.

People like M. G., Roy, Clarice, Evelyn, Arne, Isabel, Betty Anderson, and others embodied that modesty, never a false modesty that actually solicits contradiction and praise, but a genuine modesty that rises from respect for others. M. G., for example, was a sophisticated Northwestern graduate who always dressed beautifully and enjoyed reading and culture, and he devoted his life to helping farm women, women who were typically poor, poorly educated, and giving birth under challenging conditions. Yet I have uncovered no hint, written or oral, suggesting that he ever looked down upon his patients or cultivated the arrogance sometimes associated with the successful professional man. Of course, these Old Stanleyites were proud of their accomplishments and not unwilling to discuss them; M. G. told many tales of difficult or humorous births, and Roy had an equal number of stories about reaching stranded farmers in his improvised Model A snowmobile. They recounted these experiences, however, not to impress their listeners or to demean the individuals in the stories, but rather to inform or entertain—and you don't have to have grown up in Old Stanley to tell the difference.

AFTERWORD

. . . .

The Old Stanley of my memory has now largely disappeared, but before we mourn too much, let's remember that change is the essence of life, and without change there is no life. Just as I am no longer the slightly chubby boy of twelve who left Stanley in 1956, so, too, towns grow and transform. But Old Stanley's passing does impose two responsibilities on those of us who were part of it: to remember and to learn.

Some would argue that there was little to mourn over in the first place. *New York Times* reporter Chip Brown maintains that "for many years North Dakota has been a frontier—not the classic nineteenth-century kind based on American avarice and the lure of opportunity in unsettled lands, but the kind that comes afterward, when a place has been stripped bare or just forgotten because it was a hard garden that no one wanted too much to begin with, and now it has reverted to the wilderness that widens around dying towns." Such a judgment, as Jon K. Lauck has noted, comes from "correspondents sent to the region, seemingly so removed from its interior rhythms [that they] 'write like foreign correspondents.'" How easy it is for such write-and-run journalists to dismiss the region with a sneer and a boarding pass in hand.

In this book I've tried to show that Old Stanley offers us a model for how a decent small community works. It created a way of thinking about living, a distinctive set of values and virtues including resoluteness, steadfastness, devotion to community, pluck, commitment, dauntless optimism, a spirit of adventure, and modesty. Learning and living these virtues is how Old Stanley residents became natives not of a "hard garden that no one wanted too much to begin with," but of a dry place that with work, yielded much. I do not claim that this constellation of virtues was unique to Old Stanley or that they are right for everyone, only that they helped the people who lived there to build fulfilling lives and a decent society, a place whose heroes were self-sacrificing, community-oriented, modest people like M. G. Flath and Roy Edwards, rather than self-promoting, self-absorbed blowhards like Donald Trump and Rush Limbaugh. The accumulated wisdom of a place like Old Stanley should not be thrown away lightly.

Some might label Old Stanley's virtues as conservative values, but that is surely wrong, at least as the term is used today. The people of Old Stanley would find little in common with the "I've got mine," government-is-bad, culture-war fanaticism of modern right-wingers. They would reject modern conservatism's implied callousness towards less fortunate neighbors and not understand its lack of appreciation for the necessity of community and cooperation. They would disagree with its vilifying of science and education, which they revered. They loved and frequently visited national parks, an early expression of environmentalism. In these ways and more, some of Old Stanley's virtues were distinctly liberal. No, Old Stanley was a different society, with a different constellation of values and virtues. We must honor the truth in L. P. Hartley's oft-quoted line, "The past is a foreign country: they do things differently there." Indeed they do, and if we do not see the difference, if we try to force Old Stanley into today's political categories, we will not understand what it was about.

In fact Old Stanley endures, because those of us who are part of its diaspora have carried away its values and virtues and passed them to our children. Old Stanley survives in its descendants, its virtues carried to Denver and Los Angeles and Atlanta and Washington, to Boston and Brooklyn and

Paris. Adapted for new environments, mixed with other traditions and other values, its virtues persist. My children have been shaped by their mother, a Marylander, but there is some of Stanley in them, too. When my son Sam searches for the resoluteness to overcome a personal crisis, as he did during his last years of law school, he is drawing upon a deep well of family culture and habits of mind bequeathed to him by Roy and Swede, even if he doesn't know it. When my son George faces the challenges of building his own business in a hostile and competitive economy, he derives strength from a family culture enriched by the pluck of his aunts Clarice and Evelyn, even if he is unaware of its provenance. When the musical progeny of Isabel Flath develop their talent (I have no examples from my family, the gene for no talent being dominant), or when modern-day doctors, including my daughter Rebecca, an aspiring physician, follow M. G. Flath to serve whoever needs care regardless of their ability to pay, they give new life to the qualities of those natives of a dry place. And when we fail and fall short, we have a standard against which to measure our failure and plan to do better. Old Stanley and places like it continue to contribute to the American character.

Those who remain in the modern Stanley face enormous challenges, challenges fully as difficult as the ones the first homesteaders faced there. At least until we transition beyond fossil fuels, modern society needs oil to fuel SUVs in Phoenix, make petrochemicals in New Jersey, and power flights to Las Vegas, and the costs of satisfying those needs have fallen on Stanley like a tsunami. Some Stanleyites have become rich from the oil, but many others are clear losers. Perhaps an even greater challenge than this rising inequality is the fact that local residents have largely lost control of the forces shaping their lives and their town. This story is an old one, because for decades residents of the Great Plains have chafed under the influence of Minneapolis grain millers, Chicago commodity markets, Eastern-controlled railroads, and New York bankers. But their surrender of control to the oil business has been so swift and complete, and has affected everyday life so intimately, as to stun even longtime observers.

Financial rewards cannot fully compensate for what is lost when a way of life is washed away. Norman Ware studied the shoemakers and other craft producers (whom he called "industrial workers") of Massachusetts in the 1840s, ordinary people who lost their way of life in the cataclysm of the Industrial Revolution. He wrote that their resistance to factory production was sometimes interpreted as

> the result of purely temporary maladjustments. It is admitted that a temporary maladjustment lasting over one's working lifetime is sufficiently permanent for the one concerned. But it is claimed that, from the standpoint of history, the degradation suffered by the industrial worker in the early years of the Industrial Revolution can be discounted by his later prosperity. And this might be true from the calm standpoint of history if the losses and gains were of the same sort. But they were not. The losses of the industrial worker in the first half of the [nineteenth] century were not comfort losses solely, but losses, as he conceived it, of status and independence. And no comfort gains could cancel that debt.

How similar Ware's observation is to the perspective of Donny Nelson, a rancher south of the Missouri River. "I don't like what [the oil boom has] done to our communities and lifestyle," he said. "We had a good life, and now it's gone forever, or at least for my lifetime."

Old Stanley is nearly gone, and many of its traditions are maintained only by the continuing efforts of its oldest citizens, many of whom have moved elsewhere. Rapid, radical changes are inexorably reshaping the town into a new society with its own values and virtues. So the question becomes: What kind of society will it be? Will it be founded on an ethos of exploitation—grab the money and to hell with the prairie? Or will something more recognizably civic emerge?

A group called Vision West North Dakota, funded by federal and nonprofit grants and describing itself as a "community-level coalition of leaders and citizens," issued a report called the *Regional Plan for Sustainable Development*. Vision West said that the document "represents thousands of hours, miles, and voices of western North Dakota citizens' vision and plan for our future." The report documents the near collapse of basic services, both public and private, under the pressure of the oil boom. It found people's "broadest and deepest concerns" about the region's future to be child care, emergency services, housing, transportation, and water. Noting that "North Dakota has one shot at getting it right," the plan also outlines the region's "favorable factors at play," including

1. A populous [*sic*] that takes nothing for granted and still believes in hard work,

2. A positive production taxation and expenditure policy,

3. The benefit of geology that allows for extraction without long-term environmental consequence,

4. Existing communities that can accommodate growth,

5. A close-knit society in which its members know and trust one another, and

6. Unique/progressive institutions (such as the Bank of North Dakota) that offer capacity to address the present challenge.

One might legitimately ask whether this "populous" works hard enough at spelling, or whether the local geology truly "allows for extraction without long-term environmental consequence," but the most dubious resource in the list is a "close-knit society in which its members know and trust one another." Old Stanley was such a society, but now? When I called the Stanley postmaster to ask which years the long-serving Walt Poulsen was postmaster, she replied, "Oh, I wouldn't know. I've only been in Stanley a year. Everybody here is new."

Still, sometimes people surprise you, and perhaps the regional plan is a first step for the citizens of the Bakken country to reclaim some say over their future. In 2011, I stopped at Joyce's Cafe and sat in a booth. The older couple in the next booth, local farmers, were talking to a haggard-looking oilfield type slouched over a nearby table, clearly looking forward to a shave, shower, laundromat, and bed. He mentioned that he was a trucker from Texas. The couple asked if the housing collapse there was as bad as elsewhere. Expecting to hear a Sean-Hannity-type blast at Obama and how government screws everything up, instead I heard him say, "No, Texas has a lot of regulation concerning housing, and it's protected us pretty well from the meltdown." The elderly couple nodded assent. Sometimes people do surprise us. All we really know is the truth in *Mountrail County Promoter* editor Mary Kilen's words: "The changes are big and small. They may be good or bad. Whatever happens, life as you know it will not be the same."

SOURCES

· · · · · · · · · · ·

INTRODUCTION

Asker, Orba Edwards. "Orba's Stories." Ed. Jack and Edith Edwards. Unpublished mimeograph, 2000, p. 6.

Brown, Chip. "North Dakota Went Boom." *New York Times Magazine*, Jan. 31, 2013, nytimes.com.

"Continental: Bakken's giant scope underappreciated." *Oil & Gas Journal*, Feb. 16, 2011, ogj.com.

[Dalrymple, Amy], "Man Camps Create Temporary Housing Solution, but Their Future Is Unclear." *Oil Patch Dispatch*, May 6, 2012, oilpatchdispatch.areavoices.com.

Frosch, Dan. "Oil Spill in North Dakota Raises Detection Concerns." *New York Times*, Oct. 23, 2013, nytimes.com.

Gottesdiener, Laura. "A Trip to Kuwait (on the Prairie): Life Inside the Boom." *Huffington Post*, Oct. 12, 2014, huffingtonpost.com.

"Job Opening" (advertisement). *Mountrail County Promoter*, Apr. 30, 2014.

Jones, Carol Ann. "Riding the Prairie Winds." *Mountrail County Promoter*, Apr. 2, 2014.

Kilen, Mary. "Just My Opinion." *Mountrail County Promoter*, May 2, 2012.

Klimasinska, Kasia. "No Kids, No Booze, No Pets: Inside North Dakota's Largest Man Camp." *Bloomberg Business*, Feb. 12, 2013, bloomberg.com.

"Lady Roughnecks in North Dakota Man-Camps." *CNN Money*, money.cnn.com.

Lincoln Journal Star, Feb. 17, 2014.

McChesney, John. "Oil Boom Puts Strain on North Dakota Towns." *NPR*, Dec. 2, 2011, npr.org.

MacPherson, James. "In North Dakota Oil Patch, Debate over 'Man Camp' Housing Continues," *St. Paul Pioneer Press*, May 17, 2012, twincities.com.

Mason, James. "Bakken's Maximum Potential Oil Production Rate Explored." *Oil & Gas Journal*, Apr. 2, 2012, ogj.com.

Miller, Joshua Rhett. "Besieged by Oil Workers, North Dakota Town Seeks to Ban Campers." *Fox News*, Apr. 18, 2012, foxnews.com.

"More Pressure on Stanley Services." *KMOT-TV* (Minot, N.Dak.), 17 Feb. 2010.

Mufson, Steven. "North Dakota Boom Has a Price." *Washington Post*, July 18, 2012, washingtonpost.com.

North Dakota. Department of Mineral Resources. "Bakken Formation Reserve Estimates." By Julie LeFever and Lynn Helms. dmr.nd.gov.

"North Dakota Oil Spills Go Unreported." *Lincoln Journal Star*, Oct. 26, 2013.

"Oil Minister: No Change for OPEC; Shale Oil Drillers Should Be First to Scale Back." *Omaha World-Herald*, Jan. 14, 2015, omaha.com.

Omaha World-Herald, Apr. 25, 2012.

Ro, Sam. "The Middle East Has a Huge Advantage in the Global Oil Market." *Business Insider*, May 13, 2014, businessinsider.com.

"The Rules of Civility and Decent Behaviour." *George Washington's Mount Vernon*, mountvernon.org.

Santayana, George. *The Life of Reason: or the Phases of Human Progress*. Vol. 1, *Introduction and Reason in Common Sense*. New York: Charles Scribner's Sons, 1905, p. 284.

Stone, Andrea. "Oil Boom Creates Millionaires and Animosity in North Dakota." *USA Today*, Sept. 9, 2008, usatoday.com.

Sulzberger, A. G. "A Great Divide over Oil Riches." *New York Times*, Dec. 27, 2011, nytimes.com.

———. "Oil Rigs Bring Camps of Men to the Prairie." *New York Times*, Nov. 25, 2011, nytimes.com.

Tales of Mighty Mountrail: A History of Mountrail County, North Dakota. [Stanley, N.Dak.]: Mountrail County Historical Society, 1979, 1:8–11, 161.

U.S. Bureau of the Census. *Census of Population: 1950*. Vol. 1, *Number of Inhabitants*. Washington, D.C.: Government Printing Office, 1952, p. 34-14.

U.S. Bureau of the Census. "Community Facts." census.gov.

U.S. Bureau of the Census. "North Dakota: Population of Counties by Decennial Census: 1900 to 1990." Comp. Richard L. Forstall. census.gov.

U.S. Congress. *United States Statutes at Large* 26. 51st Cong., 2d sess., pt. 2, 1891, pp. 1032–35.

U.S. Energy Information Administration. "Cushing, OK WTI Spot Price FOB." eia.gov.

U.S. Energy Information Administration. "International." eia.gov/beta.

U.S. Energy Information Administration. "Petroleum and Other Liquids." eia.gov.

U.S. Energy Information Administration. "U.S. Net Imports of Crude Oil and Petroleum Products." eia.gov.

U.S. Geological Survey. "Assessment of Undiscovered Oil Resources in the Bakken and Three Forks Formations, Williston Basin Province, Montana, North Dakota, and South Dakota, 2013." Fact Sheet 2013–3013. pubs.usgs.gov.

U.S. Geological Survey. "USGS Releases New Oil and Gas Assessment for Bakken and Three Forks Formations." usgs.gov.

"Vision West ND Regional Plan Ready to Be Released." *Mountrail County Promoter*, Apr. 30, 2014.

"What's a School District to Do?" *Mountrail County Promoter*, May 16, 2012.

1. RESOLUTENESS

The primary source for this chapter was an interview I conducted with Verne Hagey in July 1978.

"Body of Thomas Scrivner Found in Abandoned Well." *Stanley Sun*, Oct. 15, 1923.

"Horrible Death Claims Life of Young Man Monday: Thom

Scrivner Found Dead in an Old Abandoned Well on the Bates Homestead." *Mountrail County Promoter*, Oct. 26, 1923.

Mandela, Nelson. *Long Walk to Freedom.* New York: Little, Brown, 1994, p. 542.

"Man Dies in Old Abandoned Well: Thomas Scrivener [*sic*] Meets Most Peculiar Accident." *Van Hook* (N.Dak.) *Reporter*, Oct. 25, 1923.

"Records of an Inquisition before Dr. A. Flath," Nov. 2, 1923. Clerk of Courts office, Mountrail County, Stanley, N.Dak.

"Register of Coroner's Inquests in the Matter of Thomas Schrivner [*sic*]," Nov. 2, 1923. Clerk of Courts office, Mountrail County, Stanley, N.Dak.

"Search for Missing Farmer Reveals Body in Bottom of Well: Thomas Scribner [*sic*], Residing Near Stanley, Thot Accident Victim." *Minot Daily News*, Oct. 24, 1923.

"Stanley Farmer Found Dead in Well." *Northwest Press* (Minot, N.Dak.), Oct. 25, 1923.

2. STEADFASTNESS

The primary sources for this chapter are interviews I conducted with Clarice Edwards Meacham on August 14, 2011, with Evelyn Edwards Hazen on August 15, 2011, with Jack Edwards on August 2, 2010, and with Irene Springan on August 23, 2010, as well as my own memories. An important source for Roy's childhood years is Orba Edwards Asker, "Orba's Stories," ed. Jack and Edith Edwards, unpublished mimeograph, 2000.

Bruns, James H. and Donald J. Bruns. *Reaching Rural America: The Evolution of Rural Free Delivery.* [Washington, D.C.: National Postal Museum, Smithsonian Institution], 1998.

"Final Payment Roll," Feb. 19, 1919. Mimeograph, National Archives and Records Administration, National Personnel Records Center, St. Louis, Mo.

Fuller, Wayne E., "RFD: The Farmers' Mail." *Timeline* 4 (Apr.–May 1987): 30–41.

Handlin, Oscar. *The Uprooted: The Epic Story of the Great Migrations That Made the American People.* Boston: Little, Brown, 1951, p. 6.

Miller, Lester F. *100 Years of Rural Free Delivery.* Alexandria, Va.: National Rural Letter Carriers' Association, 1996.

Payroll records of Stanley, N.Dak., post office, various years beginning in 1923. Mimeograph, National Archives and Records Administration, National Personnel Records Center, St. Louis, Mo.

U.S. Bureau of the Census. *United States Census of Agriculture: 1950.* Vol. 1, *Counties and State Economic Areas.* Pt. 11, *North Dakota and South Dakota.* Washington, D.C.: Government Printing Office, 1952, pp. 40, 50.

3. DEVOTION TO COMMUNITY

The primary sources for this chapter are taped interviews of Dr. M. G. Flath conducted in Stanley by the North Dakota Oral History Project on June 17, 1975, available from the State Historical Society of North Dakota; an interview of Dr. M. G. Flath conducted by his grand-nephew Tom Flath on November 26, 1984; interviews I conducted with Irene Burlingame Springan on July 30, 2011 and (by telephone) on May 2, 2014; an interview I conducted with Clarice Edwards Meacham on August 14, 2011; and an interview I conducted with Evelyn Edwards Hazen on August 15, 2011.

U.S. Bureau of the Census. *Fourteenth Census of the United States, 1920.* Manuscript population schedule, Stanley, Mountrail County, N.Dak. National Archives Microfilm Publication T625, roll 1337, sheet 1B.

State v. Flath, 228 N.W. 847 (N.Dak. 1929).

State v. Flath, 237 N.W. 792 (N.Dak. 1931).

Tales of Mighty Mountrail: A History of Mountrail County, North Dakota. [Stanley, N.Dak.]: Mountrail County Historical Society, 1979, 1:367.

4. PLUCK

The primary sources for this chapter are the interviews I
conducted with Clarice Edwards Meacham on March 3,
2009, October 3, 2010, and August 14, 2011; with Evelyn
Edwards Hazen on March 14, 2009, September 15, 2009,
and August 3, 2010; and with Betty Anderson Bergman on
September 30, 2009 and December 10, 2011.

Abbot, Carl. "Vanport." Portland State University and
Oregon Historical Society, *The Oregon Encyclopedia*,
oregonencyclopedia.org.

Carter, Susan B. *Historical Statistics of the United States:
Millennial Edition*. Cambridge, U.K.: Cambridge University
Press, 2006, table Ba4381–4390.

Clawson, Augusta H. *Shipyard Diary of a Woman Welder*.
New York: Penguin Books, 1944, pp. 31, 58–59.

Collen, Morris F., Bryan Culp, and Tom Debley. "Rosie the
Riveter's Wartime Medical Records." *Permanente Journal* 12
(Summer 2008): 84–89.

Culbertson, Lue Rayne. Interview in *Good Work, Sister!
Women Shipyard Workers of World War II: An Oral History*.
Video recording, Portland, Ore.: Northwest Women's
History Project, 2006.

Domaskin, Helen. "Helen's Memory." Unpublished
mimeograph, Mar. 25, 1995.

"18 or More by 44." *The Bo's'n's Whistle* (Portland, Ore.), Oct.
21, 1943, pp. 2–3.

Flat Top Flash (Vancouver, Wash.), Nov. 13, 1943.

Hanna, H. S. "Shipyard Health Problems." *Monthly Labor
Review* 58 (1944): 92–94.

"The History of the *St. Lo*." USS *St. Lo* (Formerly *Midway*)
CVE-63 / VC-65 Association, ussstlo.com.

Kesselman, Amy. *Fleeting Opportunities: Women Shipyard
Workers in Portland and Vancouver during World War II
and Reconversion*. Albany: State University of New York
Press, 1990, pp. 33, 51.

Kossoris, M. D. "Industrial Injuries to Women in Shipyards, 1943–1944." *Monthly Labor Review* 60 (1945): 311–15.

Lane, Frederic C. *Ships for Victory: A History of Shipbuilding under the U.S. Maritime Commission in World War II*. Baltimore, Md.: Johns Hopkins Press, 1951.

Newman, Dorothy K. "Employing Women in Shipyards." *Bulletin of the Women's Bureau* no. 192-6. Washington, D.C.: Government Printing Office, 1944, p. 83.

Rølvaag, O. E. *Giants in the Earth: A Saga of the Prairie*. Trans. by the author and Lincoln Colcord. New York: Harper, 1927.

Stevens, Marcelle Anderson. "Marcelle's Memories of Building Ships during the War Years, 1940s." Mimeograph, May 4, 1995.

Strmiska, Billie. Interviewed in *Good Work, Sister! Women Shipyard Workers of World War II: An Oral History*. Video recording, Portland, Ore.: Northwest Women's History Project, 2006.

"Vancouver Production Jumps as '18 or More' Becomes 19." *The Bo's'n's Whistle*, Nov. 25, 1943, p. 7.

"What's New in Welding." *Fortune*, Mar. 1945, pp. 15–54, 182, 184, 186, 189–90.

"Where There's Smoke . . . ." *The Bo's'n's Whistle*, Nov. 11, 1943, pp. 12–13.

5. COMMITMENT

The primary sources for this chapter are interviews I conducted with Irene Burlingame Springan on August 23, 2010, July 30, 2011, and August 16, 2013, and a typed personal history by Arne Springan, made available to me by Irene Springan. Interviews with Clarice Edwards Meacham on March 18, 2009, and Evelyn Edwards Hazen on March 14, 2009 were also helpful.

6. DAUNTLESS OPTIMISM

The primary sources for this chapter are the interviews I conducted with Clarice Edwards Meacham on January 18, 2014; with Evelyn Edwards Hazen on January 14, 2014; with Jack Edwards on January 16, 2014; and several pieces of personal correspondence from Warren Flath.

"About Us." The Sibyl Center, sibylcenter.org.

"High School Girls Now Hold Championship of State." *Mountrail County Promoter*, Mar. 22, 1917.

"Local News." *Normal Eyte* (Cedar Falls, Iowa), Mar. 2, 1910.

"Senior Class Play." *Normal Eyte*, June 11, 1910.

"Shakes to Give Play: Greek Mythological Drama to Be Presented." *Normal Eyte*, Mar. 3, 1909.

"Stanley Girls Defeat Bowbells," *Stanley Sun*, Feb. 22, 1917.

7. SPIRIT OF ADVENTURE

The primary sources for this chapter are the interviews I conducted with George Edwards on July 12, 2010, and Sam Edwards on August 9, 2010.

Corey, Elizabeth. *Bachelor Bess: The Homesteading Letters of Elizabeth Corey, 1909–1919*. Ed. Philip L. Gerber. Iowa City: University of Iowa Press, 1990.

De Smet, Pierre-Jean. *Life, Letters and Travels of Father Pierre-Jean De Smet, S. J., 1801–1873*. Vol. 1. New York: Francis P. Harper, 1905, p. 154.

"Garrison Dam." Mandan, Hidatsa, and Arikara Nation, mhanation.com.

Moulton, Gary E. ed. *The Journals of the Lewis and Clark Expedition*. Lincoln: University of Nebraska Press, 1987, 4:7–14, 11:132–34.

"1900s." Mandan, Hidatsa, and Arikara Nation, mhanation. com.

Peinado, J. Carlos and Daphne Ross. *Waterbuster*. DVD, Quechee, Vt.: Brave Boat Productions, 2006.

Stewart, Elinore Pruitt. *Letters of a Woman Homesteader*. 1914; reprint ed., Lincoln: University of Nebraska Press, 1989.

Stewart's letters were originally published serially in *The Atlantic Monthly*, and her story formed the basis for the 1979 film *Heartland*.

Tharaldson, Ardell. "Pick-Sloan and Garrison Dam." *Political Prairie Fire*, politicalprairiefire.com.

Wilson, Ron. "Garrison Dam: A Half-Century Later." *North Dakota Outdoors*, June 2003, pp. 15–19.

8. MODESTY

No additional sources were used in this chapter.

AFTERWORD

Baker, Joseph E. "Four Arguments for Regionalism." *Saturday Review of Literature*, Nov. 28, 1936, p. 4.

Brown, Chip. "North Dakota Went Boom." *New York Times Magazine*, Jan. 31, 2013, nytimes.com.

Kilen, Mary. "Just My Opinion." *Mountrail County Promoter*, May 2, 2012.

Lauck, Jon K. *The Lost Region: Toward a Revival of Midwestern History.* Iowa City: University of Iowa Press, 2013, p. 2.

Mufson, Steven. "North Dakota Boom Has a Price." *Washington Post*, July 18, 2012, washingtonpost.com.

Vision West North Dakota. *Regional Plan for Sustainable Development.* 2015 ed., pp. iii, 5, 19, visionwestnd.com.

Ware, Norman. *The Industrial Worker, 1840–1860: The Reaction of American Industrial Society to the Advance of the Industrial Revolution.* 1924; reprint ed., Chicago: Quadrangle Books, 1964, pp. x–xi.

INDEX

· · · · · · · · · ·

Sacagawea (Lemhi Shoshone Indian), 153

Sanish, N.Dak., 117, 119, 162–63

Sather, Ben, 34–37, 42–44

Scrivner, Anna Niemi, 43

Scrivner, Awrey, 31

Scrivner, "Old Man," 31–32, 36, 42

Scrivner, Tom, x, 29–44

Sexual harassment, 24, 75–76, 98–99

Shulkin's Confectioners, 50

Sibyl Center, 3, 13

Slim (welder), 102–3

Socialism and communism, 13, 64, 71, 100. *See also* Communist Party

Social Security, 64

Spokane, Wash., 89

Sports championships, 22, 139–40, 146

Springan, Arne, 82, 106, 109, 112–33, 153–54, 159, 162, 166–67, 173

Springan, Betty, 113–14, 123–24, 128–31, 133

Springan, Dawn, 131

Springan, Gail, 124–25

Springan, Henry ("Flash"), 23, 60, 82, 84–85, 133

Springan, Irene Burlingame, 74, 112–33, 135–36, 153–54, 162, 167

Springan, Mae, 130

Springan, Mark, 124–25

Springan, Paul, 116, 127, 130–31

Springan's Furniture, 1, 59, 84–85

Stanley, N.Dak.: beginnings and early development, 20–23; and oil boom, 3–4, 6–7, 11; population, 3, 21. *See also* Mountrail County, N.Dak.; "Old Stanley"; specific businesses and institutions

Stanley High School, 139–40

Stanley Police Department, 7

Stanley Sun, 22–23, 41, 59, 61–64, 68

Stewart, Elinore Pruitt, 149

Stewart, W. R., 47

St. Lo (escort carrier), 101

Strmiska, Billie, 100–101

Sweet Shoppe, 117

Syrians, 16, 84

Telephone service, 22, 29, 37, 54

Tiisto, Oddie, 56

Tiisto family (Mountrail County), 9

Tioga, N.Dak., 4–5, 9

The Uprooted, 69

U.S. Postal Service. *See* Rural mail delivery

U.S. Weather Bureau, 64–65